Labor

Labor

ANDREW HEROD

polity

First published in 2018 by Polity Press

Polity Press
65 Bridge Street
Cambridge CB2 1UR, UK

Polity Press
101 Station Landing, Suite 300
Medford, MA 02155, USA

ISBN-13: 978-0-7456-6386-9
ISBN-13: 978-0-7456-6387-6(pb)

A catalogue record for this book is available from the British Library.

Library of Congress Cataloging-in-Publication Data

Names: Herod, Andrew, 1964- author.
Title: Labor / Andrew Herod.
Description: Cambridge, UK ; Malden, MA : Polity Press, 2017. | Series: Resources | Includes bibliographical references and index.
Identifiers: LCCN 2017015282 (print) | LCCN 2017030878 (ebook) | ISBN 9781509524112 (Mobi) | ISBN 9781509524129 (Epub) | ISBN 9780745663869 (hardback) | ISBN 9780745663876 (pbk.)
Subjects: LCSH: Labor.
Classification: LCC HD4901 (ebook) | LCC HD4901 .H435 2017 (print) | DDC 331--dc23
LC record available at https://lccn.loc.gov/2017015282

Typeset in 10.5 on 13pt Scala by Servis Filmsetting Ltd, Stockport, Cheshire
Printed and bound in the UK by Clays Ltd, St Ives PLC

For further information on Polity, visit our website: politybooks.com

Contents

Figures and Tables

FIGURES

TABLES

Acknowledgments

I would like to thank several people who have helped me with preparing this book. First, Louise Knight, who approached me several years ago now to contribute to Polity's Resources series. I would also like to thank Nekane Galdos for helping to shepherd the project along. My friend Bradon Ellem granted permission to use the map in Figure 5.1. Kavita Pandit helped me track down some of the demographic data in Chapter 2. I would like to thank Adam Wrigley for help with some of the data collection. Adam participated in an innovative program at the University of Georgia in which high school seniors are paired with faculty members so that they can learn a little about how research is conducted. Adam did a sterling job locating some of the data used in the book. Thanks, also, to two anonymous reviewers for some helpful comments. Finally, I dedicate this book to my dad, John Clement Herod (1931–2013).

Introduction

Labor is the ultimate source of all wealth. Without the exertion of human labor, the world's natural resources, like iron ore, oil, lumber, and land, cannot be used and neither can the services that make the modern world run be provided. This capacity to transform the planet around us through work is central to understanding the dynamics shaping contemporary economic, political, social, and biotic life. In thinking about labor as a resource, however, we must ponder not only labor itself but also the labor processes within which individuals are embroiled and which shape how they interact with one another as they refashion the natural world for human use. This is because the way in which working people cooperate (or not) with one another and with their employers changes how they themselves behave as a resource. Because they have the capability to contest and change the conditions under which they are employed, then, it is important to think of workers both as *objects* of analysis and also as sentient *subjects* who can alter their own situations through proactively thought-out economic and political actions. This capacity makes labor fundamentally different from any other resource.

In this book I provide a broad-ranging look at labor as a resource. In the first chapter I outline how labor's sentience makes it a resource unlike others. In so doing, I consider how labor is deeply geographically embedded in particular places but is also distributed unevenly across the economic landscape. This spatial aspect of labor's nature is more than simply an interesting facet of working people's existence. Rather, it is a deeply constitutive element of their lives, for much of their behavior is shaped by their need to come to terms with the differentially developed geography of the contemporary world and how they are tied to specific places. This has signal implications for labor's behavior as a resource. For instance, many workers are trapped in particular communities for a variety of reasons – an inability to sell their homes so that they can move elsewhere, a psychological attachment to particular places that hold important resonances for them, or the fact that they lack the legal status to move elsewhere. This can reduce their ability to negotiate better terms of employment because they have few options but to take whatever wages are being offered locally. On the other hand, for those workers who can overcome this geographical fixity, the ability to migrate elsewhere may provide significant opportunities for bettering their lives. At the same time, though, it may be fraught with challenges – they may not be welcome in the new communities in which they settle, the promises of a better life may not materialize, or they may have to live in the economic shadows because they do not have the correct legal documents to work out in the open. How they negotiate the tensions between staying in one place and seeking to move elsewhere, then, is an important component of their lives.

Having explored several aspects of what it means to think about labor as a resource, how it is different from other resources, and how labor's geographical constitution affects how it behaves, I look in Chapter 2 at labor as a global resource. In particular, I detail some of the historical migrations which have led to the current distribution of people across the planet, together with how *in situ* demographic processes like differential rates of population growth in various parts of the globe, have resulted in the present distribution of working people across the planet. For instance, the booming oil economies of the Persian Gulf are reliant upon the labor of millions of migrant workers coming from places like India, Pakistan, and Bangladesh, while South Africa's economy sucks in young men from across the continent to work in the gold mines of the Witwatersrand. Likewise, the migration of hundreds of millions of peasants from the countryside to the nation's industrial zones has helped make China's economy the world's second largest. Concurrently, the global population explosion that has occurred since the mid-twentieth century has fundamentally reconfigured labor's distribution across the planet – since 1950, world population has tripled in size (from about 2.5 billion to nearly 7.5 billion), with the bulk of this growth occurring in the Global South. At the same time, however, regional phenomena like the AIDS crisis in Africa are having tremendous impacts upon labor markets and what this means for economic development.

Chapters 3 and 4 explore two phenomena that are dramatically impacting working people across the planet, albeit in different ways: globalization and the growth of work precarity. In the first of these chapters, I examine globalization's impact on working people, focusing especially on the geography of foreign direct investment (FDI) and how this both has allowed employers to play workers in different parts of the globe against one another but has also brought these workers into greater contact with one another, thereby creating improved opportunities for them to challenge their employers' actions. A key feature of such globalization has been the growth of numerous global production networks (GPNs) through which commodities pass as they are manufactured, with these networks tying workers in different parts of the globe together through how the labor process is structured. I also investigate what I term global destruction networks (GDNs) through which discarded commodities like old electronics pass as they are broken up so that their constituent elements can be recovered for reuse as inputs into new products. These GDNs often link people in the Global North, who discard old commodities, with workers in the Global South, who take them apart. For its part, Chapter 4 details some of the ways in which work and life have become more precarious for millions across the planet. Much of this precarity is the result of neoliberal policies which have made labor markets more "flexible." Hence, many workers who previously enjoyed full-time, long-term employment now find themselves reduced to part-time, short-term arrangements, while millions of others find themselves with few opportunities for generating paid income except to participate in the "gig economy" as they flit from job to job with companies like Uber and Lyft, food delivery services like Bite Squad and Caviar, and many others. At the same time, new technologies threaten to replace workers across the planet with machines. Meanwhile, for millions of farmers in places like Africa's Sahel, global climate change is threatening harvests and leading many to migrate to Europe to escape their ever more precarious lives.

Much as Chapters 3 and 4 operate as a pair, so do Chapters 5 and 6. Specifically, they investigate the supposed transition from an "Old Economy" dominated by manual labor and "dirty" jobs to a "New Economy" of knowledge work and mental labor. Although US management consultant Peter Drucker first used the term "knowledge worker" in 1969, it became popularized in the 1990s when growing numbers of writers began suggesting that a new way of organizing economies and work was emerging, one that would have tremendous implications for labor. This

New Economy, they averred, would be characterized by "a world in which people work with their brains instead of their hands" and would be "at least as different from what came before it as the industrial age was from its agricultural predecessor," such that "its emergence can only be described as a revolution" (Browning and Reiss 1998: 105). The future of work, then, would be one in which the dreary, labor-intensive toil of the past would be replaced by clean and rewarding technology-aided manufacturing and service work in which workforces would be empowered and liberated from drudgery. In Chapter 5, in light of this assertion, I examine work in three economic sectors – iron ore mining in Australia, cocoa plantation work in West Africa, and labor in the global fishing industry. As I show, there is little in these sectors which looks much like the imagined emancipatory work that is supposed to characterize twenty-first-century economies. This perhaps is understandable, given how these are activities usually associated with Old Economy work. In Chapter 6, on the other hand, I detail work in several sectors that are often viewed as emblematic of the New Economy, including hi-tech manufacturing and various types of service work. Significantly, despite the utopian language with which the work that is imagined to define the New Economy is often portrayed, much of this labor is also little different from the dirty work of the Old Economy. Rather than there being a historical transition from an Old to a New Economy, then, I suggest that the emancipated work enjoyed by many "knowledge workers" is in fact built upon a foundation of drudge work carried out by millions of highly exploited workers.

The final chapter surveys some of the ways in which workers challenge their positions within various labor markets and labor processes and how they seek to self-organize to improve their lots. This includes forming labor unions and other types of organizations, as well as engaging in activities like occupying factories and running them on their own rather than under the direction of various managers. In this regard, the chapter focuses upon how working people act as the subjects of their own histories and geographies and are not merely resources used by others.

CHAPTER ONE

A Resource Unlike Any Other

Labor has long been thought of as a key resource shaping how economies work. Classical economists like Adam Smith and David Ricardo conceived of resources as falling into one of three categories: natural resources (minerals, flora and fauna, water, land, and so forth, which are provided by the Earth but which can be transformed into finished products for use by people); the capital stock, by which they meant humanly produced goods (machinery, tools, buildings, etc.) used to produce other goods; and labor, which they understood to be the human effort (both physical and mental) required to transform the natural world. In such a view, "labor" came to be understood, by implication, as referring also to the people doing this work. For his part, Karl Marx, who was, like Smith and Ricardo, a classical economist, viewed resources pretty much entirely in terms of labor, suggesting that a society's key resources – what he called the "productive forces" – were labor (workers who controlled their own capacity to labor, or their "labor power"), the subjects of labor (the commodities produced when human labor transforms the natural world), and the instruments of labor (machinery, tools, etc.). Within this framework, Marx viewed capital as simply "congealed labor," that is to say, the surplus value extracted from workers in the capitalist production process. For their part, the neoclassical economists of the late nineteenth century saw resources in terms of the factors of production (land, labor, and capital) outlined by the classical economists, but they interpreted them in slightly different ways. Whereas for classical economists the worth of a commodity is a reflection of the amount of labor time involved in producing it (an approach known as the "labor theory of value"), for neoclassical economists the "value" of a commodity is not a reflection of the amount of labor incorporated within it but, rather, its price – i.e., what someone is willing to pay for it. Put another way, for classical economists a commodity's *value* (the amount of labor time incorporated within it) and its *price* (the amount of money that someone will pay to purchase it) are different measures of its worth; for neoclassical economists, on the other hand, a commodity's value is measured in terms of its price and not the labor time incorporated in it. In other words, for neoclassical economists price = value = price, whereas for classical economists price ≠ value ≠ price (for more on these differences, see Wolff and Resnick 1987). More recently, some economists have suggested that, in addition to the three broad sets of resources of land, labor, and capital, there is a fourth: human capital (i.e., skills and education). Some others have argued that energy should be considered a separate resource. Within such discussions, different economists have argued over which of these various resources should be considered the most important.

Below, I am not going to argue for which resource is most important – though my personal feeling is that labor is the *primus inter pares* resource, for it is the only one capable of proactively transforming its own conditions of use. Rather, what I want to do is to suggest that labor is a resource unlike any other for one specific reason: it is sentient. Without doubt, some might aver that certain other resources also have sentience – I am thinking here of fauna like cattle, which can be used both as implements of labor (drawing plows, for instance) and as inputs into other production processes (their meat can be used for food, their skins for making leather, bones can be used as fertilizer, etc.). However, although both cattle and humans are obviously sentient, there is an important difference in their sentience: while animals clearly have consciousness, what distinguishes them from humans is the latter's capacity for proactive planning – as Marx (1867[1967]: 174) put it, "what distinguishes the worst architect from the best of bees is this, that the architect raises his structure in imagination before he erects it in reality." This human sentience means that we must view labor in two ways: as an *object of analysis* (i.e., as a resource that awaits use by a potential employer) and as a *subject in analysis* (i.e., as a resource capable of altering its own conditions of existence and of challenging how it is used by others).

In considering how labor is a resource unlike others, it is perhaps helpful to turn to the writing of two geographers, Michael Storper and Richard Walker (1983), who sought to theorize how labor markets operate. Although their analysis focuses upon commodities rather than resources, their approach is nonetheless useful for our purposes here. Specifically, in discussing why labor is a commodity unlike others, they argued (1983: 4) that what they called "true commodities" (raw materials and manufactured goods) can generally "be industrially produced, purchased at a consistent price and standard quality, owned outright and employed in a strictly technical manner, however and whenever the owner wishes," although they recognized that this is "somewhat less true for plants and animals than inanimate objects." As a result, these types of commodities can be understood largely in terms of performance versus cost – how much must I pay for a good of such-and-such quality? Labor, though, is quite different.

Thus, although it takes a commodity form in that it can be bought and sold, labor "is not a true commodity; it is a *pseudo-commodity*."[1] This is because its sentience means that it has the capacity to shape any system of production within which it is imbricated and to challenge the conditions under which its use is bought and sold. Labor's sentience also means that it has the ability to play active roles in reproducing itself biologically and socially and to control, at least in part, the situations in which it does so (unlike resources like cattle and coal). Hence, Storper and Walker asserted, "Labour is fundamentally different from every other production input because people are conscious subjects of production."

In drawing inspiration from Storper and Walker, we can identify several important dimensions of labor as a resource, all of which relate to labor's sentience. The first relates to "conditions of purchase." For sure, like other commodities, labor is purchased on the market. In this regard, it takes on a commodity form and has a price for purchase – the wage rate. However, as Storper and Walker indicated (1983: 5):

> [T]he price of labour is normally more complex than just wages [for the] conditions of purchase for labour include such things as safety and health, security and regularity of employment, prospects for advancement and fringe benefits, etc. Such considerations never enter the picture where true commodities, such as forklifts, are concerned. Yet they often matter more than wages to workers, and they certainly invoke real costs for whoever must bear the burden of their presence or absence, workers or capitalists.

The point, then, is that although true commodities like iron ore or oil and the pseudo-commodity of labor both have a price, and that price will vary based upon the quantity and the quality of the resource being purchased, only one of these actually cares about the nature of what is being purchased and how. Coal or gold or a cow do not concern themselves about their purchase price nor their quality as a resource. Labor, on the other hand, usually cares deeply about both its purchase price (i.e., its wage) and how much it is expected to work for that wage (i.e., its quality to its employer).

The second aspect of labor that makes it a different type of commodity/resource from others is its "performance capacity." By this, Storper and Walker are referring to its cost compared to its productivity, rather than just to its purchase price. Thus, although two workers may be hired for the same wage, one may be of higher quality (i.e., more productive) than the other. Whereas other types of resources may also vary in quality – some sources of gold are purer than others, for instance – in such cases, their performance capacity can more easily be determined ahead of time by scientific or other analysis. The performance capacity of workers, however, is often not known ahead of time because there are many factors shaping their quality, including their level of technical expertise, as well as more nebulous characteristics like their creativity and persistence in the face of new challenges, their patience and emotional stability, their mood on any given day, whether or not they are sick, and so forth, all of which will affect how productive they are. Moreover, this productivity can vary over the course of their work day – they may be more productive in the morning but then start to tire around lunchtime until they have something to eat, at which point they perk up until it is time to end work. It can also vary over the course of their working life. Furthermore, unlike the quality of other resources, many of these characteristics are also under the direct control of workers themselves, who may be upset with their boss or spouse on one day but not another, which can affect how hard and/or carefully they tackle their work.

Third, the issue of labor control is quite different from that of how other resources are controlled and used. Unlike natural resources like oil or iron ore, which do not consciously "resist" being incorporated into the production process and whose "resistance" can usually be determined ahead of time through a knowledge of chemistry and physics, or resources like animals, which may have some capacity to resist how they are used but can simply be slaughtered if they prove too troublesome, workers can dramatically contest the conditions under which they are incorporated into the production process, even after their labor time has been purchased. Thus, even if it were possible to know ahead of time all the potential issues involved in using labor as a resource (a worker's skill level, emotional stability, degree of patience, etc.), once it is paid for, workers' actual performance capacity is not the same as their potential performance capacity because they have the ability consciously to defy their employer's control over them and to limit their work effort. As Storper and Walker (1983: 6) put it:

> Workers, unlike machines [and, I would add, other types of resources], must willingly engage their capacity for work and have the power to resist their use by [their employer]. The intensity, continuity and quality of work that can be got out of workers, with what degree of supervision, monitoring and punishment, is of fundamental importance to the employer. The usual term for the problem of eliciting performance is "labour control."

As they point out, however, such control is a double-edged sword, for "it is not enough that workers follow the employer's orders; they must actively participate in production." Consequently, "capitalist production contains a fundamental contradiction between the need for labour control and the need for thinking, creative working people."

A fourth way in which labor is unlike other resources relates to its ability to reproduce itself, socially and biologically, on a daily and generational basis, largely beyond the control of whomever might hire it. Certainly, workers who toil for someone else (as opposed to those who are self-employed) spend part of their time during any given work day, and during their working lives, under the watchful eye of their employer or, in the case of peasant farmers, perhaps under the gaze of their feudal lord. In this regard, how they reproduce themselves socially and biologically can be controlled, to a degree, by their employer, who can set rules concerning how they must dress, with whom they can socialize, how they are to behave in the workplace, and even when they may go to the toilet. However, once workers leave the workplace, they are usually beyond their employer's gaze and so the latter does not have such direct control over them. They are thus much freer to behave as they wish and to shape their own mode of self-reproduction. Consequently, "a substantial part of labour's reproduction takes place in the home and community, beyond the reach of the employer ... In other words, workers cannot be industrially produced as are true commodities" (Storper and Walker 1983: 6).

Although Storper and Walker's analysis – of how labor is a pseudo-commodity unlike other commodities – is helpful for thinking about how it is a resource unlike others, I would add a fifth consideration: not only is labor involved in the production of goods and services, but it is also an active and self-conscious consumer of them. Other resources do not share this characteristic. This means that labor is an important element not just in the production of profit but also in its realization and thus for the circulation of capital and the continuation of the capitalist system. This provides another way in which we can think of labor as having a dual character unlike other resources – it is involved in both the active production of goods and also their active consumption. In this way, the factors outlined above that play an important role in shaping labor as a resource in production also shape it as consumer. Hence, labor as consumer makes conscious decisions about what to consume in ways similar to how it decides how to behave in the production process. Equally, labor as consumer is distributed in a geographically uneven manner across the economic landscape in much the same way as is labor as a producer, while its consumption patterns are differentially shaped by its geographical location (goods available in some places may not be available elsewhere).

Having outlined why labor is a resource quite unlike others, in the next two sections I turn to the issue of how we might think of labor geographically, for any understanding of a resource must fundamentally be about its distribution across the Earth's surface. As we shall see in Chapter 2, labor is more available in some places than in others. This is important because, much as the spatial distribution of other resources has had important implications for patterns of local economic and political development, so, too, does the availability of labor shape the possibilities for development in different places. How labor is spatially differentiated and embedded in the economic landscape, then, is an important aspect of understanding it as a resource. This means that we must think both historically (how does labor as a resource develop over time?) and also geographically (how does it develop over and in space?). In thinking about labor geographically, however, it is important to distinguish two important ways of doing so: seeing labor as both an object of analysis and a subject in analysis. The difference between these two relates fundamentally to the matter of labor's sentience and its capacity to adapt itself to the geographical contexts within which it finds itself.

Labor as Object

One way to think about labor as a resource that is deeply geographically differentiated and unevenly spread across the landscape (i.e., that has both a history *and* a geography) is to approach the matter from the viewpoint of those who hire workers and/or wish to catalogue their characteristics in some way. In this perspective, labor is an *object* of analysis. Working people are seen not so much as sentient beings capable of making decisions that can shape how the economic landscape is produced but, rather, either in the abstract terms of how they fit into the locational decision-making process of their employers, or simply in reductionist terms of various statistics – how many workers are located in Place A compared to Place B, how much they are paid, what types of work they do, and so forth. Such a view has a long history in locational and economic analysis for, as Aronowitz (1990: 171, emphasis added) has argued, "[t]he history of capitalism has, typically, been written as a series of narratives unified by the themes of accumulation." These narratives have been marked by a focus upon

> mercantile and imperialist interests seeking fresh sources of investment; the scientific and technological revolutions that have driven growth; international rivalries over territory and labor supplies and the multitude of conflicts among fractions of capital that take political forms, such as the struggles for power among capital's personifications or wars ... *In these accounts, workers enter the theater of history as abstract labor, factors of production, dependent variables in the grand narratives of crisis and renewal.*

One of the first social scientists to develop a systematic theorization of how labor shapes the economic landscape's form was geographer Alfred Weber (brother of German sociologist Max Weber). In his 1909 treatise *Über den Standort der Industrie* (published in English in 1929 as *Theory of the Location of Industries*), Weber explored how labor costs, transportation costs, and what he called "agglomeration forces" (the advantages that accrue to an individual firm from locating close to others) affect the locational decisions of independent, single-plant firms. However, his approach focused only upon understanding how employers choose between different groups of workers when making locational decisions, with the result that working people were reduced merely to the characteristics that might be of interest to their potential employers – wages, skill levels, militancy, gender, and so forth. In the process, much of their humanity was lost.

Other fields of study have similarly treated labor in the terms outlined by Aronowitz. Thus, labor economics, the one subfield of economics devoted specifically to the study of labor, views workers largely as abstract players in its models of how labor markets operate, reducing them to numbers within complex mathematical formulae. For its part, human capital theory, which focuses upon labor-force characteristics to suggest that, by developing their skills, working people may find better employment, marginalizes working people as sentient social beings through its very terminology – referring to their efforts as being about developing their "human capital" sees labor as just another aspect of capital. Equally, the product cycle model, commonly used in the 1960s, 1970s, and 1980s, explained the emergence of the new international division of labor as resulting from corporations relocating production from Global North countries, where labor is expensive, to Global

South ones, where it is cheaper. However, while it considered labor an important shaper of investment, it did so only from the point of view of how corporate decision-makers choose between competing groups of potential employees across the planet.

Whereas the above approaches have largely relied upon neoclassical economics as an explanatory framework, many Marxists have also tended to treat labor in rather passive terms, as little more than variable capital – that is to say, they have similarly conceptualized labor as an object rather than a subject in analysis. On the one hand, this is not surprising. Marx, after all, was primarily concerned with understanding the dynamics of capital accumulation and the extraction of surplus value from labor and, although he occasionally made reference to how workers resist exploitation by their employers, this was not his main focus of analysis. Nevertheless, it has meant that many Marxists have treated labor in much the same way as have neoclassical analysts, viewing working people simply in terms of how potential employers choose between them when making hiring decisions. For instance, in their book *The Capitalist Imperative*, Storper and Walker (1989), although arguing for the centrality of labor as a resource in shaping the location of particular economic activities, nonetheless were principally concerned with showing how the supply and demand of particular types of labor shapes how capitalists decide where to locate their firms. Likewise, Richard Peet's (1983) analysis of the geography of US class struggle showed how, in the period just after World War II, much manufacturing industry left the old industrial heartlands of the northeast, where labor was stronger, and relocated to the southern and southwestern states, where it was cheaper and largely non-union. However, like Storper and Walker, Peet's approach largely viewed things through the eyes of the firms that were making these locational decisions, not the workers themselves. As such, he also viewed labor in descriptive terms of it being an object of analysis.

For sure, the above is an extremely foreshortened account of how some writers have thought about labor. However, the examples do highlight a particular perspective on labor as a resource. In suggesting that the approaches outlined above view labor as an object of analysis, one stripped of its essential humanity, such that workers are viewed in terms little different from sheaves of wheat, deposits of oil, or expanses of lumber, I do not want to leave the impression that such an approach cannot be useful. In fact, it can be very helpful when considering how employers make decisions about which resources and how much of them – in this case, workers – to use in different places. What I do want to argue, though, is that it is also important to view labor from the perspective of itself, as a self-aware social and geographical actor, for it is this dual aspect, labor as both a resource to be used by others and a resource capable of shaping its own historical and geographical destiny, that is central to why it is quite unlike other resources.

Labor as Subject

Whereas the approaches highlighted above have viewed labor as an object of analysis, here I contrast them to perspectives which have seen working people as multidimensional socio-spatial actors capable of modifying the conditions of their own existence – that is to say, which views them as the subjects of their own histories and geographies and thus considers labor as a *subject* of analysis. Almost by definition this excludes most (all?) approaches within the field of economics, for these tend to disconnect working people from those social aspects of their lives that cannot be quantified and reduce them to abstract numbers for inclusion in various mathematical models of the economy and decision-making. However, there are some fields of study in which labor has been viewed as the subject of its own history and, more recently, its own geography too. One of the most important English-speaking writers of the past half century to place labor at the center of its own history is E.P. Thompson, whose 1963 book *The Making of the English Working Class* explored the life-experiences of eighteenth- and nineteenth-century working people as they came to see themselves as having a common set of experiences which shaped how they behaved. Thompson's key argument was that the English working class was present at its own creation – as he put it (1963: 194): "The working class made itself as much as it was made." His book is important because it focused upon concrete labor (i.e., workers as people rather than as abstractions) and telling "history from below" (Aronowitz 1990: 175).

Arguably, the other academic field where labor has been placed at the center of its own existence is that of geography and, specifically, the field of "labor geography" (for a brief summary, see Herod 2001; Lier 2007; Rutherford 2010). Emerging in the early 1990s, this field has argued that labor's geographical situation shapes its potential for economic and political praxis, but that, concomitantly, working people's economic and political praxis shapes their geographical situation. Labor geographers, then, have adopted an approach which views labor as making its own geographies but not under the conditions of its own choosing, to paraphrase Marx's famous quote.[2] This is because, in order to reproduce itself on a daily and generational basis, labor must ensure that the landscape is made in some ways and not others – as a landscape of employment rather than unemployment, for instance. Working people's efforts to create a landscape with a configuration sufficient to allow them to live and reproduce themselves – what Harvey (1982) has called creating a "spatial fix" – require them to make sure that they have work close to where they live, a roof over their head, places for their children to go to school, and so forth, which, of course, shapes how they act as a resource. For instance, to ensure that they have access to income-providing work, working people may decide that they have to migrate to other places. However, they may be significantly constrained in seeking to create such a new spatial fix for themselves by family ties, an inability to sell their home, and/or an unwillingness to leave behind their assets and the burial places of their ancestors. Consequently, they may instead decide that it makes more sense to try to bring jobs to their own communities, in which case they might ally themselves with local pro-growth groups or temper their militancy for fear that asking for wages that are too high could scare off new investment. While, then, workers may be constrained by their spatial embeddedness in particular places, through their actions to create new spatial fixes appropriate to their condition they play active roles in reconfiguring the economic landscape. There is, in other words, a dialectic at play – the possibilities of their economic and political action are shaped by the spatial contexts within which they live but, simultaneously, their economic and political action reshapes those spatial contexts.

In thinking of labor as a resource that seeks to reshape the economic landscape as part of its effort to further its own self-reproduction, five important points come to the fore. First, such activity means that the making of the economic landscape is not the sole prerogative of those who employ working people (i.e., labor as object); it is carried out also by those working people themselves (i.e., labor as subject). In this regard, then, labor is not simply a passive "factor" of location, but an active socio-spatial player. Second, recognizing that working people have geographical objectives they seek to obtain to ensure their self-reproduction allows us to place the economic landscape's production at the very heart of their social praxis, even if working people do not necessarily see themselves as engaged in making the economic landscape. Third, the fact that different groups of working people may have quite different visions for how they want the economic landscape within which they live to be structured means that we should not view "labor" as an undifferentiated mass. This has implications for how we think of labor as a resource. Hence, different groups of working people may react differently to similar situations, based upon their history, their sense of being tied to a particular place, the type of political environment within which they live (liberal democracy versus authoritarian state), and so forth. This means that whereas a lack of employment in their community may lead some to migrate elsewhere, for others the same situation may encourage them to stay put and to try to develop opportunities for employment locally.

Fourth, the fact that working people often seek to impose different types of spatial fixes at different times means that, whatever else they may be, processes of class formation and inter- and intra-class relations are fundamentally geographical in nature and must be understood as such. This affects how labor acts as a resource, because how working people think of themselves – whether as individuals or as part of a broader socioeconomic class – will shape how they behave in interactions with their employers. Finally, the fact that working people may make linkages with one another across space as part of their political praxis means that they are involved in reconfiguring the geographical scales at which they self-organize – when working people make linkages with those elsewhere in order to help them in a labor dispute, for instance, they are quite literally rescaling the dispute from being a local one to a nonlocal one. Likewise, working people may link communities in different regions or countries by migrating from one to another and then sending money home in the form of remittances to support their families (see Chapter 2). Making such linkages is important because they can shape working people's structural capacities to resist being exploited as a resource – as we will see in Chapter 7, developing such trans-spatial connections makes it more difficult for employers to play workers in one place against those elsewhere, which shapes how these employers can use labor as a resource.

Summary

There are two key issues to take away from this discussion. First, labor is a resource quite unlike others because of its sentience and capacity for proactive social, economic, and political praxis, both when it comes to producing goods and services and also with regard to consuming them. Because of this sentience, labor is both an object of analysis as well as a subject in analysis. No other resource has this double nature. Second, labor is a resource that is deeply geographically differentiated (like most other resources), but it also, through its agency, has the capacity to alter the conditions of its own existence, with this capacity being fundamentally shaped by its geographical situatedness. Put another way, labor as object is clearly unevenly spread across the economic landscape, but labor as subject has the capacity to transform, to a greater or lesser degree, how that economic landscape is made, together with its own distribution across it. As a resource, then, labor, as object and subject, is both prisoner of its own history and geography and also a maker of that history and geography. Understanding labor's capacity for historico-geographical praxis and how this is shaped by working people's geographical circumstances is important, therefore, for understanding labor as a resource.

CHAPTER TWO

Labor in Global Context

Why is labor distributed across the globe today in the way in which it is? Answering this question requires looking at two interconnected processes: labor's migration from place to place and differential population growth in particular places. In terms of the first, approximately one in seven people globally is a migrant (in 2014 there were some 232 million international migrants and 740 million internal ones: IOM 2014a). In terms of the second, populations in different parts of the globe have historically increased at varying rates because they have been affected by distinct sets of demographic conditions at various times. Together, these two processes determine the size of the potential labor force in any particular place and, therefore, often what kinds of work can be done in them. In what follows, I explore each of these sets of processes and how they intersect with one another.

Moving On

All our ancestors originated in East Africa. Some remained. Others left. The earliest remains of modern humans (*Homo sapiens sapiens*) are found in Ethiopia and date to nearly 200,000 years ago. Beginning about 100,000 years ago, humans started to venture out of Africa and have been colonizing different parts of the globe ever since. Given this, we could start the story of how labor as a resource has populated various parts of the globe a very long time ago. However, for purposes of providing a manageable overview of labor's geo-graphical distribution across the planet, I will confine my account to migrations of the past 500 years or so, as this is the period when the "Old World" of Europe, Africa, and Asia began to be increasingly linked to the "New World" of the Americas. When considering human migration as a shaper of the geography of labor as a resource, however, it is important to think about how this has occurred at dif-ferent spatial scales: within countries, within continents, and between continents. Clearly, it would be a mammoth undertaking to provide an account of every single migration in every part of the world, even if we just consider the past half millennium. Consequently, here I will outline some of the most significant.

Rural to urban migration
In terms of migrations within countries, arguably the most significant of these historically has been the migra-tion of people from the countryside to cities. Although rural to urban migration is as old as cities themselves, it gathered significant speed with the coming of the indus-trial revolution in the eighteenth century. In particular, industrialization – first in Britain and, later, in other European countries, together with countries like the United States – required that large numbers of workers become available to labor in the mines and factories around which the new urban metropolises were developing. As a result, whereas in 1700 about 13.3% of the population of England and Wales lived in towns of at least 10,000 people, by 1800 that figure had jumped to 20.3% and by 1850 it was 40.8% – the equivalent figures for Scotland are 5.3%, 17.3%, and 32.0%, respectively (Lynch 2003: 30). In terms of larger cities, in 1801 about a tenth of the people of England and Wales were living in cities of 100,000 people

or more, but by 1900 this figure had jumped to about 30% (Davis 1965: 5). In France, where industrialization took hold later, the proportion of the population living in towns grew more sluggishly and even recorded a decrease in the late eighteenth century – it was 9.2% in 1700, dropped to 8.8% by 1800, but then grew to 14.5% by 1850 (Lynch 2003: 30).[1] Between 1846 and 1911, 79% of the growth in France's urban population was due to migration (Davis 1965: 12). In the United States, the proportion of the population that was urban grew from 5.1% in 1790 to 15.3% in 1850 and 39.7% in 1900 (Haines 1994: Table 2). Across the planet, then, the first century and a half of industrialization led to significant urbanization of the world's population as people migrated to towns and cities in search of work (see Table 2.1). In turn, this migration affected rural areas, which often then suffered labor shortages. For instance, in Britain the peak number of males employed in agriculture (1.8 million) was reached in 1851 but declined to just 0.5 million by 1961. In France, the rural population fell from 26.8 million in 1846 to 20.8 million in 1926 and 17.2 million in 1962, even though the overall national population grew significantly and fertility rates were higher in rural areas than in urban ones (Davis 1965: 9, 13).

Table 2.1 Percentage of global population living in urban areas, 1800–1950		
	Urban areas of 20,000 or more	Urban areas of 100,000 or more
1800	2.4	1.7
1850	4.3	2.3
1900	9.2	5.5
1950	20.9	13.1

Source: Wittman 1973: 115

By encouraging the rural to urban migration of people to work in the new factories, industrialization not only affected European cities and workers; it also had consequences for workers elsewhere. In particular, it increasingly linked working people in places like Britain with those in other parts of the world, for, as agricultural laborers left the countryside, someone still had to grow the food upon which the nascent industrial labor forces were coming to rely. That someone was often farmers and agricultural laborers in the colonies or elsewhere, such that British workers' daily bread was increasingly made using wheat grown in places like Canada, Australia, India, the United States, and Russia's Crimea (see Table 2.2). With growing demand to feed industrial workers, other types of food from abroad was also increasingly imported into Britain. For instance, 156,490 head of live cattle were shipped from the United States in 1880, in addition to 724,272 hundredweight of fresh beef. As a way to extend access to such food, British investors established thirty-seven Anglo–US cattle companies on the Great Plains between 1879 and 1900 (Brayer 1949: 91–92). Other countries likewise became big food suppliers for Britain, including Argentina (where British capital was heavily involved in expanding both the beef industry and the railways, which brought beef from the interior to the port of Buenos Aires) and the British colony

Table 2.2 British wheat consumption, in million cwts. (112 lbs), 1880–1909		
Yearly averages	Net home grown	Net imports
1880–89	40.0	75.2
1890–99	32.6	93.8
1900–09	27.5	110.2

Source: Schooling 1911: 51

of New Zealand, which, beginning with the first lamb ship-ments in 1882, exported meat, butter, and other agricultural goods to feed British workers. Overall, these developments meant that between 1851 and 1910 the amount of meat imported into Britain from abroad rose from a weight of 0.01 pounds per person annually to 28 pounds (Eliott 1916: 4). Industrialization in Britain, then, not only drove the rural to urban migration of working people within the country; it also had significant impacts upon land use and working people's activities across the world.

Rural to urban migration continues to shape the geo-graphical distribution of the planet's labor forces. Every year, millions of migrants – especially in the Global South – leave their homes in the countryside in search of a better life. Arguably, this phenomenon has been most evident in China, where between 1979 and 2010 the urban popula-tion grew some 480 million (to 666 million) and where, as Chan (2013: 981) notes, even if "only half of that increase was due to migration, the volume of rural–urban migration in such a short period is likely the largest in human his-tory." As a result, about 44% of Chinese now live in urban areas (Gautreaux 2013). Indeed, this rural–urban migration has been so great a shaper of where working people live that the twentieth century can fairly be called the century of humanity's urbanization – while in 1900 approximately 220 million people lived in urban areas, in 2000 2.84 bil-lion people did so (UNFPA 2007: 7). By 2014, 3.9 billion (54% of the global population) were urbanites, and this figure is expected to rise to 6.4 billion (about two-thirds of global population) by 2050 (United Nations 2014: 1), with the bulk of this increase due to migration. The result has been that, whereas in 1850 there were only two cities world-wide (London and Paris) with a population of more than 1 million people, by 2000 there were nearly four hundred.

By 2030, estimates suggest, there will even be forty-one "mega-cities," each with more than 10 million inhabitants (United Nations 2014: 1). However, although humanity has been rapidly urbanizing over the past century, it is important to recognize that it has done so in a geographically uneven manner: in the Americas about 80% of the population are urban dwellers, in Europe the figure is 73%, and in Africa and Asia the figures are 40% and 48% respectively. These two latter continents, though, are changing very quickly, such that by 2050 their urban populations are expected to be 56% and 64% of their overall populations (United Nations 2014: 1). Meanwhile, the world's rural population has grown fairly slowly since 1950 and is expected to reach its peak soon: presently amounting to about 3.4 billion people (nearly 90% of whom are in Africa and Asia), the planet's rural population is expected to decline to about 3.2 billion people by 2050 (United Nations 2014: 1).

Intracontinental migration
A second migratory process that has been important historically and continues to be so is long-distance migration within continents. For instance, since the 1880s such migration has been key to the South African economy's development, as its gold and diamond mines have swallowed up millions of migrant workers from across southern Africa. Whereas in 1899 there were 90,000 black miners working in South African gold mines (Harington et al. 2001: 66), in 1920 there were 200,000 (Crush 1986: 27). By 1940 that number had jumped to 360,000 and in 1986 there were 477,000, though this number declined with labor retrenchments in the 1990s (Harington et al. 2001: 67–68). Use of such migrant labor provided benefits for both the mine owners and the various governments involved. Because many of the workers were foreign and

had few rights, mine owners could pay them poorly, largely without challenge from the government, which saw the availability of cheap labor as important for exploiting the country's natural resources. Foreign miners could also easily be deported should they either be no longer needed or become too troublesome by trying to unionize. Meanwhile, for the governments of the countries from whence they came, the export of young men to work in South Africa's mines not only provided a source of foreign exchange, as the governments could tax miners' remittances sent home to support their families, but these governments also made money because mine owners paid them commissions for securing this labor. Sending migrants to the mines could also reduce unemployment in miners' home countries.

The Gulf oil states have also come to rely upon migrant labor, in this case people from Bangladesh, Pakistan, India, the Philippines, and elsewhere. Using the *kafala* system, individuals and companies in Saudi Arabia, Qatar, Bahrain, and elsewhere have sponsored millions of foreigners to work drilling oil, doing construction, or working as maids (see Table 2.3). For these migrants, many have better job prospects than in their home countries, while Gulf citizens enjoy higher living conditions because the work done by this foreign labor has allowed their economies to expand more rapidly than would otherwise have been possible, thanks to a lack of local labor. However, there are also challenges that arise from these countries' reliance upon migrant workers – the workers have few rights and are often highly exploited, whereas many of the native-born population feel alienated in their own countries because they are a minority population within them. The significant imbalance in the sex ratios of the populations also creates issues, with many young men unable to find female company (e.g., see Figure 2.1). However, the phenomenon of labor migration

Table 2.3 Native and foreign components of GCC member states' labor forces, 1975–2015 ('000s)

		1975	1985	1990	1999	2008	2015
Saudi Arabia	Nationals	1,027	1,440	1,934	3,173	4,173	5,591
	Foreigners	773	2,662	2,878	4,003	4,282	6,321
	Total	1,800	4,102	4,812	7,176	8,455	11,912
	% foreigners	42.9	64.9	59.8	55.8	50.6	53.1
Kuwait	Nationals	92	126	118	221	351	302
	Foreigners	213	544	731	1,005	1,742	1,341
	Total	305	670	849	1,226	2,093	1,643
	% foreigners	69.8	81.2	86.1	82.0	83.2	81.6
Bahrain	Nationals	46	73	127	113	139	158
	Foreigners	30	101	132	194	458	567
	Total	76	174	259	307	597	725
	% foreigners	39.5	58.0	51.0	63.2	76.7	78.2
Oman	Nationals	137	167	189	312	276	394
	Foreigners	71	300	442	503	809	1,546
	Total	208	467	631	815	1,169	1,941
	% foreigners	34.1	64.2	70.0	61.7	74.6	79.6

Table 2.3 (Continued)

		1975	1985	1990	1999	2008	2015
Qatar	Nationals	13	18	21	36	72	94
	Foreigners	54	156	230	244	1,193	1,448
	Total	67	174	251	280	1,265	1,542
	% foreigners	80.6	89.7	91.6	87.1	94.3	93.9
UAE	Nationals	45	72	96	124	455	763
	Foreigners	252	612	805	1,165	2,588	4,324
	Total	297	684	901	1,289	3,043	5,087
	% foreigners	84.8	89.5	89.3	90.4	85.0	85.0
TOTALS	Nationals	1,360	1,896	2,485	3,979	5,466	7,302
	Foreigners	1,393	4,375	5,218	7,114	11,072	15,547
	Total	2,753	6,271	7,703	11,093	16,538	22,849
	% foreigners	50.6	69.8	67.7	64.1	66.9	68.0

Source: for 1975–2008 data, see Baldwin-Edwards 2011: 9; data for 2015 from: Kingdom of Saudi Arabia General Authority for Statistics; Kuwait Central Statistical Bureau (figures are for 2013); Bahrain Labour Market Indicators, Labour Market Regulatory Authority, 4th Q 2015; *Sultanate of Oman Statistical Yearbook for 2015* (figures are for 2014); Qatar Statistics Authority Quarterly Labor Force Survey 2013, 4th Q 2013 (figures are for 2013); United Arab Emirates data are from the CIA *World Factbook*, with figures for number of nationals and foreign workers calculated by the author.

Note: "2008" data for Kuwait are for 2009; "2008" data for Bahrain are for 2010, employed only; % foreigner figures for UAE, 2015, are estimates

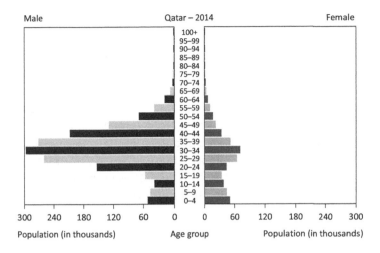

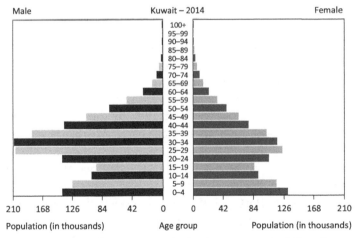

Figure 2.1.   Population pyramids for Qatar (2014) and Kuwait (2014)

Source: www.indexmundi.com/qatar/age_structure.html; www.indexmundi.com/kuwait/age_structure.html

can have significant benefits for the sending countries. Between them, foreign workers in Gulf Cooperation Council (GCC) states sent some US$80 billion worth of remittances to their home countries in 2014 (*Daily Sabah* 2016).[2] In 2013, the Philippines government estimated that some 10.2 million Filipinos/Filipinas (about 10% of the country's population) worked or lived abroad (many in GCC countries), while official remittances in 2011 were placed at a value of US$20.1 billion (Magtulis 2012; Commission on Filipinos Overseas 2013). Indeed, these remittances are so important that the government has presented migrant workers as "national heroes."

Intercontinental migration

Finally, in considering migration's role in shaping how labor is distributed across the globe, it is crucial to account for intercontinental labor movement. Here, too, there are potentially hundreds of mass migrations – both voluntary and involuntary – upon which we could focus. One of the most significant – as it played a vital role in linking the Old World with the New – was undoubtedly the trans-Atlantic slave trade. Although there had been African slaves in Europe going back to at least Roman times, the start of the modern trade can be dated to 1444, when two Portuguese ships brought twelve slaves from Cabo Branco (in today's Mauritania) to Portugal. Within a few years, the Portuguese were bringing larger numbers of Africans to Europe and by 1452 slaves were being sent to the island of Madeira, off Portugal's coast, to work in the sugar fields recently established. The trade was facilitated by Pope Nicholas V's issue of a papal bull (*Dum Diversas*) authorizing the Portuguese to enslave any non-Christian. As European voyages to the New World expanded, so did the slave trade. Christopher Columbus took several Native Americans back to Europe as slaves, while Amerigo Vespucci and Alonso de Hojeda transported hundreds from the coast of South America to Spain. Whereas the first African slaves taken to the Americas left from Europe, not Africa, by the 1530s slaves were being imported directly from Africa. Over time, the numbers increased, reaching about 30,000 per year by the 1690s and 85,000 per year by the late eighteenth century. Such was the trade's magnitude and its impacts upon the supply of labor to the Americas that, by 1820, for every European who had landed in the New World there were approximately four Africans, and four out of every five women who had crossed the Atlantic were African (Gilda Lehrman Institute of American History n.d.). Although a lack of complete records makes it difficult to tell with certainty, most historians believe that 9–11 million slaves made this journey, though the total numbers affected were much higher – many died crossing the Atlantic and an additional 9–11 million are believed to have died in Africa, the result of being forced to march from the interior to the slave castles along the coast, fighting slave traders, or awaiting transportation while in captivity. More than 90% of African slaves were imported into the Caribbean and South America and only about 6% went to British North America, although by 1825 one-quarter of all people of African descent in the New World were living in the United States. Slaves arriving in the New World came from different parts of Africa – those from Angola overwhelmingly went to Brazil, whereas those from Senegal mostly went to North America.

The practice of African slavery did not just link the New World with Africa, though. Beginning in about the eighth century, the Arabs were also involved in trading African slaves, although their trade was mostly focused upon bringing slaves captured south of the Sahara to North Africa and also exporting them across the Indian Ocean

(India's present-day Siddis are their descendants) or even to the Far East (Harris 1971). Although it is difficult to give exact figures, estimates of the number of slaves exported between the ninth and nineteenth centuries range from 8 to 17 million. Whatever the precise number, not only did such practices provide significant labor in the societies to which the slaves were transported, but it also reduced an important source of labor availability in Africa itself, which substantially hindered the continent's economic development. Indeed, as Congolese historian Elikia M'bokolo (1998) has put it:

> [Africa] was bled of its human resources via all possible routes. Across the Sahara, through the Red Sea, from the Indian Ocean ports and across the Atlantic. At least ten centuries of slavery for the benefit of the Muslim countries (from the ninth to the nineteenth). Then more than four centuries ... of a regular slave trade to build the Americas and the prosperity of the Christian states of Europe. The figures, even where hotly disputed, make your head spin. Four million slaves exported via the Red Sea, another four million through the Swahili ports of the Indian Ocean, perhaps as many as nine million along the trans-Saharan caravan route, and eleven to twenty million (depending on the author) across the Atlantic Ocean.

Although the African slave trade has clearly been one of the most significant migrations – involuntary though it was – of labor from one part of the world to another in the past five centuries, it is, of course, not the only one to have helped shape labor's present distribution. Migration from other parts of the world has also dramatically impacted labor supply and labor market dynamics in different places. For instance, between 1820 and 1900, 18.7 million legal immigrants obtained lawful permanent resident status in the United States, with more than 90% of these coming from Europe, nearly 2% from Asia, 7% from other parts of the Americas, and negligible numbers from Africa and Oceania (Department of Homeland Security 2014: 6). During the twentieth century, an additional 46 million legal immigrants arrived, though there were significant differences in their origins, with fewer migrants coming from Europe (about 46%) and more from other parts of the globe – some 17% from Asia, about 34% from other parts of the Americas, 0.5% from Oceania, and 1.4% from Africa (Department of Homeland Security 2014: Table 2). Significantly, not only have such immigrants' origins changed over time, but the levels of immigration into the United States have done so too. Thus, peak immigration was reached during the first decade of the twentieth century, while periods of war, economic depression, and restrictions on immigration (such as the 1924 Immigration Act, designed to limit immigration from southern and eastern Europe, Africa, the Middle East, and Asia) saw immigration levels fall: between 1900 and 1909, 8.2 million legal immigrants obtained permanent resident status; between 1930 and 1939 not quite 700,000 did so.

Although immigration has clearly provided the United States with large numbers of workers and a cheap source of labor over the past two centuries, other parts of the Americas have also been important destinations for millions of migrants. In the late nineteenth century, Japanese began arriving in South America, especially Peru and Brazil, creating ongoing links between these countries; Brazil today has the largest population of Japanese descent outside Japan (about 1.5 million people), while Brazilians form the largest non-Asian population in Japan (about 312,000). Japanese workers also migrated to Hawaii to work in the sugarcane and pineapple fields. Indeed, such was their level of migration that, whereas in 1853 indigenous Hawaiians

made up 97% of the population, by 1923 the largest population in Hawaii was Japanese (Library of Congress n.d.). At
the same time that many Japanese were heading overseas,
Korean workers were migrating to Japan after their country
became a Japanese colony – by 1930, an estimated 300,000
had done so (Min 1992: 10). Meanwhile, thousands of
Chinese migrated to Hawaii as well as to the US mainland,
working in California's gold fields, building the nation's railroads, and finding labor opportunities in other occupations.
Equally, when restrictions placed upon the transportation of
convicts began to dry up Australia's labor supply in the mid-
nineteenth century, employers looked to new labor supplies.
Consequently, between 1851 and 1861 more than 600,000
migrants arrived in Australia. Although the majority were
from Britain and Ireland, there were also 60,000 from
Continental Europe, 10,000 from the United States, just
over 5,000 from New Zealand and the South Pacific, and
42,000 from China (Government of New South Wales
2010). Indeed, Chinese immigration into Australia was such
that in the early 1860s one in nine men in Australia was
Chinese (Gittins 1981). Meanwhile, between 1830 and 1920,
about 3.5 million indentured workers left India and went
to places as far afield as the Caribbean, Mauritius, and
Réunion (to work sugar plantations), Malaysia, South Africa,
and Dutch, French, and British colonies in South America.
Portuguese cocoa producers in São Tomé and Príncipe
likewise relied upon imported labor, securing access to
131,000 Angolan contract laborers between 1880 and 1910
(Ishemo 1995: 164). In the British Empire, from 1850 to
1910 more than 150,000 indentured laborers departed one
of its regions for another every decade, with the result that,
by 1914, as many as 14,000 migrants had left British West
Africa for Guyana, some 64,000 Chinese had gone to South
Africa, and 63,000 Pacific Islanders had gone to Australia's

Queensland (Northrup 1999: Table 5.1). Importantly, these figures do not include the nonindentured migrants who also left one part of the Empire for another – for example, the 119,000 Chinese who left Hong Kong between 1854 and 1880 and went to Australia, or the 1.6 million Indians who went to Malaysia between 1844 and 1910. Nor do they include those who left British imperial possessions for non-British parts of the world – for example, the 5,000 Hong Kong Chinese who left for Cuba between 1856 and 1858, or the 224,355 migrants who left India for the United States between 1854 and 1880 (Northrup 1999: Tables 5.2 and 5.3).

Much of the nineteenth century's international labor migration, then, involved people leaving Europe and heading to other parts of the globe (see Figure 2.2). Large numbers of British and Irish migrants went to North America, Australia, and South Africa. Many French people left for places like Algeria, after France conquered it in the

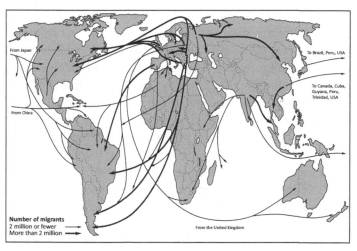

Figure 2.2   Global voluntary migrations, 1815–1914

Source: Hirst and Thompson 2001: 25

1830s. About 1.4 million Portuguese left for Brazil during the nineteenth and early twentieth centuries, seeking economic opportunity and escape from political repression at home (Segal 1993: 16). There was also significant labor migration within the Global South, as noted above – Indian indentured laborers heading to places like the Caribbean, for instance. Finally, there was also some migration from the Global South to the Global North, though this was fairly limited – even in the United States, the Global North country that historically has probably received the largest number of migrants from the Global South, between 1820 and 1950 only about 2.67 million legal permanent migrants came from Asia, Central America, the Caribbean, South America, and Africa (about 6.4% of the total immigration into the United States during this time) (Department of Homeland Security 2014). During more recent years, however, global labor migration patterns have changed significantly. For the most part, migration from the Global North to the Global South is not as significant as in the past and the migration streams linking these two parts of the globe now largely comprise people leaving the Global South and heading north (see Figure 2.3). Hence, of the nearly 232 million international migrants in 2013, 59% lived in the Global North, of whom about 60% came from the Global South. Although Europe hosted the largest number (72 million), Asia was close behind, at 71 million. They were followed by North America (53 million), Africa (19 million), Latin America and the Caribbean (9 million), and Oceania (8 million). Of the 53 million international migrants who came to the Global North between 1990 and 2013, 42 million (78%) were born in the South (United Nations 2013: 1–2). In the United States, between 1950 and 2010 some 27,116,000 legal permanent immigrants (constituting 74.7% of all such immigrants during this period) came from Asia,

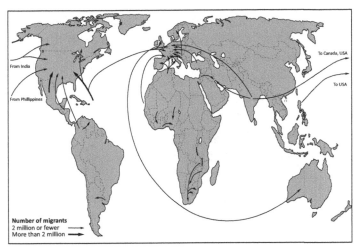

Figure 2.3   Global voluntary migrations, 1945–1980

Source: Hirst and Thompson 2001: 25

Central America, the Caribbean, South America, and Africa (Department of Homeland Security 2014).

With regard to the European Union (EU), the other main Global North destination, in 2015 34.3 million inhabitants were born beyond its borders, while an additional 18.5 million people born in one EU state were living in another. Depending upon how countries are classified, six of the ten most significant contributors to this non-EU migrant population were Global South countries (see Figure 2.4), and new EU citizens are coming overwhelmingly from Global South countries. In 2015, about 2.7 million migrants and refugees (the bulk from Syria, Afghanistan, and Iraq) arrived in the EU, while 727,200 people newly acquired EU citizenship, of whom 31% came from Africa, 14% from North and South America, and 21% from Asia (Eurostat 2017). Given the ongoing economic crises in Latin America, growing numbers of migrants from Brazil, Argentina, Chile, and Mexico

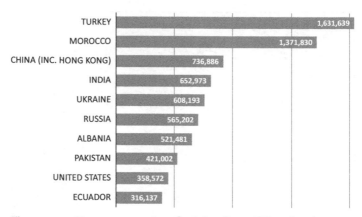

Figure 2.4   Top ten countries of origin of non-EU nationals residing in the EU, 2015

Source: European Commission 2015

have been arriving in recent years, facilitated by policies in some countries that allow people from former colonies to gain nationality after a period of time – Spain, for example, permits people from its former empire in Latin America to gain nationality after residing in the country for two years. This, of course, also gives them legal access to other parts of the EU. However, with the recent economic predicaments in Spain, Portugal, and Italy, whereas the dominant migration stream since the late 1990s had been from Latin America to the EU, beginning in 2010 this stream reversed itself, illustrating just how sensitive to economic conditions such long-distance labor migration can be – in 2012, more than 181,000 EU nationals emigrated to Latin American and Caribbean countries but only 119,000 people from there moved to the EU, a 68% drop since the historic highs of 2007 (IOM 2015).

Although there is significant South–North migration, then, another very important set of contemporary labor migration flows is occurring between countries within the Global South, a phenomenon that has been growing in recent decades. In 2013, for example, 86% of the 96 million international migrants then living in the Global South came from other Global South countries, with the single largest South–South flow being that of the estimated 3.2 million Bangladeshis living in India (United Nations 2013: 1; 2012: 3). Another noteworthy case of South–South migration involves China, which has become both a source of migrants to Global South countries and a recipient of migrants from them. As a result of reforms that began in the 1970s and reduced barriers to leaving the country, by 1990 about 4.1 million people born in China were living overseas and by 2000 about 5.5 million were doing so. Although China actually has a fairly low emigration rate, the 9.3 million Chinese who had left the country by 2013 nevertheless made it the world's fourth-largest source of international migrants. China's emigration stream has changed over time, though, and Chinese emigrants are increasingly moving to more industrialized economies; the proportion who migrated to Global North countries grew from 53.4% in 1990 to 58.6% in 2013, with a concomitant reduction in emigration to the Global South. However, there are significant differences between migrants who are highly skilled and/or wealthy and those who are relatively unskilled. For example, in 2010–11 about 1.8% of China's highly educated workers left the country, a rate that was five times that of the general emigrant population. The migration streams of wealthy and highly skilled are also quite different from those of unskilled emigrants. Whereas wealthy and highly skilled migrants mostly go to Global North countries, those with fewer skills migrate to both Global North and Global South countries (Xiang 2016: 2–3). This migration has taken place largely through one of two means: migrants leave either because they are tied collectively to specific development projects, or they leave as individuals – of the 527,000 who migrated in 2013, 271,000 were project-tied and 256,000 were individual migrants. Whereas in the former category migrants are hired by Chinese companies working on projects around the world, in the latter they are supposed to be hired by overseas companies who pay them a salary (though it sometimes does not work out this way). Of these latter, about 40% work in manufacturing, 25% in construction, and 15% in agriculture, fishing, and forestry. Fewer than 0.5% work in white-collar occupations. In the early 1980s, many went to oil-producing states like Kuwait and Iraq. However, after the First Gulf War the migration streams changed, and between 1991 and 2008 about 70% headed to East and Southeast Asia, 20% went to Africa, 5% to Europe, 3% to Australia and New Zealand, and the remainder to other destinations, including US Pacific dependencies like Saipan and Guam. Much of this migration is linked to patterns of foreign direct investment, with significant migration to oil-rich countries like Nigeria and Sudan. With the global financial collapse in 2008, together with the 2011 Japanese earthquake, however, migration patterns have changed again – the number of Chinese going to Japan dropped from 68,188 in 2007 to a mere 1,923 in 2012, while migration to South Korea (Japan's former colony) was suspended between 2008 and 2011 because the countries' governments could not agree on recruitment procedures in light of the impacts of the financial crisis on Korea's economy (Xiang 2016: 8–10).

As Chinese migrants head overseas, a growing number of foreigners are heading to China. Given Chinese investments in Africa, it is perhaps not surprising that growing numbers of Africans are calling China home. Although the data are imprecise, most analysts estimate that about

20,000 Africans are living in southern China, although one Chinese source has reported 100,000 in the city of Guangzhou alone (IOM 2014b: 4). The largest group of African migrants are Nigerians, followed by those from Cameroon, Côte d'Ivoire, Gambia, Ghana, Guinea, Mali, the Democratic Republic of Congo, Senegal, and Tanzania. The majority of these migrants appear to be traders seeking either to buy cheap Chinese goods to resell at home or to establish trading linkages for African raw materials like oil and minerals. Some are students, who are attracted by government scholarships and who stay on after graduation, while others are English teachers. Finally, there is a growing number of informal workers, who often labor as domestic servants, deliverers of home-cooked food, or guides and agents for African business people. Other groups are also increasingly coming to China, often illegally. Thus, despite its massive overall population, recent declines in Chinese fertility rates related to things like its one-child policy have encouraged labor shortages in some places. In 2004, for instance, some 2 million job vacancies went unfilled in coastal southeastern China and within a year labor shortages had spread northwards into the Yangtze River and north coastal region. With shortages for certain types of workers, southern Chinese employers are reportedly smuggling into China workers from countries like Vietnam, many of whom have Chinese ancestry and some language skills and so can blend more easily into local populations. There is also growing immigration of entrepreneurs and industrialists from South Korea into China's northern cities and irregular flows of refugees from North Korea who often have few skills and end up in manual labor and/or service-type activities (Skeldon 2011).

A frequent pattern with such South–South migration is for individuals to move first from the countryside to cities and then overseas. Another important element is the movement of labor from one Global South country to another, with the goal being to use this second country as a jumping-off point for movement to a Global North one. For instance, in recent years growing numbers of migrants from Asia and sub-Saharan Africa have been arriving in Latin America in the hope of making it into the United States or Canada, especially as EU immigration policies have been tightened. Some of these migrants, though, end up staying in Latin American and Caribbean countries because their economies have been expanding and because many find them less xenophobic/racist than the United States. In 2011, for example, about 40,000 Chinese were believed to be living in Suriname, whereas the Chinese immigrant community in Brazil is the sixth-largest foreign community in the country, the result of some growing strategic sectors of the economy, like textiles, agriculture, and information technology communication, turning to Chinese immigrants for their labor needs.

Growing in Place

Although migration, both historically and contemporaneously, has significantly shaped how labor is distributed geographically, it is also important to examine the *in situ* processes at work across the planet. Presently, about 60% of global population is concentrated in Asia, with sizeable numbers also in Africa (16%) and Europe (10%). However, this distribution has changed over time – whereas in 1900 Europe housed about 24.7% of total world population and Africa 8.1%, it is estimated that by the end of the twenty-first century these figures will be 6.8% and 23.6% respectively (see Figure 2.5). It has particularly changed since the end of World War II – while in 1950, 68% of the

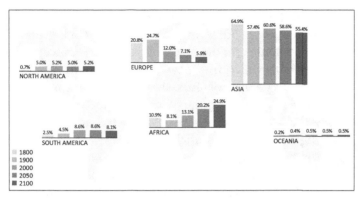

Figure 2.5   The changing proportion of the world's population living in different continents, 1800–2100

Source: for years 1800, 1900, 2000, 2050, see United Nations 1998; for year 2100, see United Nations 2004

world population lived in Global South countries, by 1975 it was 74%, by 2000 it was 84%, and it is estimated that by 2050 it could be as high as 87% (United Nations 2015a). This shift is even more remarkable given that between 1950 and 2015 the Global North was a net receiver of international migrants while the Global South was a net sender, with the volume of net migration generally increasing over time (United Nations 2015b: 6). Part of the reason for this shift in the balance of population has to do with the fact that Global South populations have experienced a sizeable demographic explosion and part has to do with the fact that fertility rates in most Global North countries – especially in Europe – have dropped drastically. This is leading to an aging of the population, which is creating concerns that there will not be sufficient new workers coming into the labor force to support those about to leave it and those already retired. Significantly, for thinking about labor as a resource, different countries have sought to address this issue in different ways. In Europe, many have turned to immigrants as a source of new workers, looking either to former colonies (in the case of Britain and France) or to countries with which they have longstanding linkages (as with Germany's use of Turkish migrants, beginning in the 1960s). In Japan, on the other hand, the cultural preference for a more homogeneous society has meant that ensuring there will be sufficient workers was historically solved either by tapping those who were not traditionally employed (women and the elderly) or through developing labor-saving technologies so that fewer workers are needed. In recent years, though, the situation has reached the point where the government and employers are now having to consider turning to immigrants. Consequently, the number of immigrant workers has been growing, with most coming from countries either that are geographically close to Japan or that have historic connections to it because they served as destinations for Japanese migrants in earlier times (e.g., Brazil and Peru).

One of the most important processes shaping the distribution of populations – and, therefore, potential labor forces – has been that of industrialization, beginning in the late eighteenth century. Between about 8,000 BCE (the invention of agriculture and the founding of the first cities) and 1750, global population increased by approximately 1.0% per annum; between 1750 and 1950, however, European industrialization occasioned a considerable increase in the rate at which the population expanded – it grew about 1.3% per year. The wealth that industrialization brought to those societies experiencing it, then, encouraged significant population growth, largely by reducing the death rate – there was more money to spend on public health, people could afford better diets, and so forth. Of course, given that industrialization proceeded in a geographically uneven manner, its impacts upon the global population have been uneven. In 1801, the population of the United Kingdom and Ireland was some 16.3 million; by 1851, it had grown to 27.5 million and by 1901 to 41.6 million – increases that helped Europe's share of the global population rise from 20.8% in 1800 to 24.7% by 1900. Importantly, though, as detailed above, industrialization in Britain did not impact just British workers but also workers in other parts of the world, linking them in new and different ways. Hence, because the British industrial revolution was largely based upon the textile industry's mechanization, British manufacturers needed two things: raw cotton and new markets to soak up the surplus products that could not be sold in Britain. They found both of these in India. Consequently, Indians were increasingly turned into producers of raw cotton and consumers of British cotton goods – between 1814 and 1835, exports of British cotton goods to India increased fifty-one-fold, while exports of raw cotton from India mushroomed, from 34.5 million pounds in 1846, to 204 million pounds in 1860, to 615 million pounds in 1866 (for more details, see Herod 2009: 121–125). In turn, many Indian manufacturers were run out of business, especially after the British introduced various tax schemes aimed at destroying the indigenous textile industry that had existed prior to India becoming a British colony. Transformations in Indian food production, as more land was devoted to cotton production, and in the economy more generally, as the textile industry was eviscerated, had important demographic consequences for India. Between 1871 and 1921, for instance, overall population growth was relatively slow, and in some regions it was actually negative (Kumar 1983: 490). We can but ponder what the population of India – currently the world's second-most populous nation – might have been today without such impacts.

If industrialization led to one demographic revolution in the late eighteenth and early/mid-nineteenth century, another – that associated with the widespread use of anti-biotics – began in the post-World War II era and resulted in an even greater explosion of population: from 2.5 billion in 1950 to 6.2 billion in 2000 (a growth rate of about 2.9% per year) and 7.5 billion in 2017. Whereas the demographic effects of the industrial revolution largely impacted the Global North in the eighteenth, nineteenth, and early twentieth centuries, the "medical revolution" has largely impacted the Global South, leading to a significant reduction in mortality (especially of children, who are, of course, the next generation of workers) and thus to significant population growth. However, given that many parts of the Global South have yet to industrialize – India has a rural population of 857 million and China's is 635 million (69% and 47% of their respective total populations) (United Nations 2014: 1) – it is entirely possible that rates of growth in some parts may drop appreciably as we progress into the twenty-first century, as happened in Europe. Certainly, China's birth rate has declined dramatically since the 1980s, from about 21 births per 1,000 people in 1985 to only 12.1 per 1,000 in 2013. Whether a result of its one-child policy or other factors like industrialization, the slowing of China's population growth rate is having significant implications for the economy's ability to secure labor. With fewer young people being born, the population is aging, yet there will be fewer workers entering the labor market in the coming decades to support those who have retired from it. This phenomenon will affect China not only internally but also externally, as worker shortages may drive up the cost of labor and thus affect Chinese goods' competitiveness on the world market.[3]

Across the globe, population growth is impacting different regions' labor markets in distinct ways. Although

Africa has the fastest growing population, Asia adds the greatest number of people every year. If present growth trends continue, by 2050 an additional 2.5 billion people will have been added to the world population, nearly 90% of whom will be found in Asia and Africa. Twelve countries alone (India, China, the United States, Indonesia, Nigeria, Pakistan, Brazil, Democratic Republic of the Congo, Ethiopia, Philippines, Mexico, and Egypt), only one of them in the Global North, are expected to account for half the projected population growth. Significantly, although populations in Africa and Asia are still relatively rural compared to others, with only 40% and 48% of people respectively currently living in urban areas, both are urbanizing faster than are other regions and are projected to have 56% and 64% of their populations living in urban areas by 2050, with clear implications for labor markets in both rural and urban areas. Indeed, India, China, and Nigeria alone are expected to account for about 37% of the growth in the world's urban population – India is projected to add 404 million urban dwellers, China 292 million, and Nigeria 212 million (United Nations 2014: 1). Furthermore, whereas by the early 2020s the populations of India and China are both expected to reach about 1.4 billion people, India's population will soon outstrip that of China, growing to 1.5 billion in 2030 and 1.7 billion in 2050, while China's is predicted to decrease after the 2030s. This will have significant impacts upon the global economy, both because India's labor force will grow significantly and because India is an English-speaking country, which brings it certain advantages when its call centers compete for contracts servicing customers in the United States and Britain (see Chapter 6). For its part, Nigeria (also an English-speaking country) will have the third-largest population by about 2050 (it presently has the seventh-largest). By the middle of the twenty-first century, of the six most populous countries, only one (the United States) will be in the Global North and the other five (China, India, Indonesia, Nigeria, and Pakistan) will be Global South ones (United Nations 2015b: 4).

Lastly, one phenomenon that is having a dramatic impact upon demographic processes, and thus labor as a resource, is Africa's HIV/AIDS crisis. Although the history of HIV/AIDS in Africa is complicated and contested, as of 2011 some 23.5 million people in sub-Saharan Africa were infected (about 70% of all cases worldwide) and nearly 5% of the population aged 15–49 – that is to say, the working-age population – was infected (the equivalent figure for the world as a whole was about 0.8%). Furthermore, people infected with the strain of the virus that is most prevalent in Africa (HIV-1, subtype C) tend to progress to full-blown AIDS more quickly than do people infected with subtype A, the most common variant in the Americas and Europe. As a result of both the high levels of infection and the virulent nature of the particular strain, combined with poorer access to healthcare, the HIV/AIDS epidemic is having a much more significant impact on the labor force in Africa than it is elsewhere. Specifically, it is affecting labor forces through seven "key impact channels" (Drimie 2002), these being: (1) a reduced labor force; (2) lower labor productivity through illness and absenteeism (of both the infected people and those who care for them); (3) cost pressures for companies through benefit payments and replacing workers; (4) lower income, as employees bear some of the AIDS-related costs; (5) lower population translating into lower expenditure across the economy as a whole, with the knock-on effects thereof; (6) increased private-sector demand for health services; and (7) higher government expenditure on health services. The impacts of these are being felt differently across the continent, however. In South Africa, for example, one of the industries most affected has been mining, which is a mainstay of the economy and is largely staffed by migrant labor; with workers living away from home for long periods of time, active sex industries exist in and around mining camps (Corno and de Walque 2012). In Zimbabwe, however, with nearly 70% of the population living in rural areas and dependent upon smallholder agriculture, the HIV/AIDS crisis is having its most significant impacts in the countryside – Stover and Bollinger (1999) documented that the crisis was leading to a decline in cultivated acreage due to a shortage of labor and that households with a member who dies from HIV/AIDS saw average reductions in maize production of 61%, in cotton of 47%, in vegetables of 49%, in groundnuts of 37%, and in cattle owned of 29%. In Tanzania, in households containing an AIDS patient, an average of 29% of family labor was spent on taking care of them, and if two family members were involved in such caring activities, the total labor loss was 43%. Firms employing large numbers of people infected with HIV/AIDS have also seen significant labor turnover and/or the loss of sizeable numbers of skilled workers.

Summary

The geographic distribution of population is clearly an important aspect of understanding labor as a resource. There are two components to this: migration and *in situ* growth (or decline). Although these have been discussed somewhat in isolation from one another, it is important to recognize that they are actually tightly interconnected. Hence, people who migrate from one part of the globe to another for work tend to be in their prime childrearing years – in 2015, 72% of all international migrants were in the 20–64 age range, compared to 58% of the total global population (United Nations 2016: 13). Given that they are collectively younger than host populations and frequently come from societies where large families are the norm, migrants usually have a higher fertility rate than do those of the societies to which they are migrating, with the result that the immigration of working-aged people can appreciably increase the rate at which the destination society's population grows. The gendered nature of much labor migration can also impact family structure and social dynamics in both the migrants' origin and their destination countries. Thus, husbands might leave their wives and children behind as they look for work, thereby shaping both fertility rates in their home countries and also patterns of social development as children grow up with fathers often absent. In the case of the destination countries, having large numbers of single men can impact both the migrants themselves and the receiving societies – of the nearly 40,000 Chinese in Australia in 1861, for example, only 11 were women, which resulted in many men never finding a bride or marrying a non-Chinese woman (Australian Government 2015). At the same time, the overwhelmingly male nature of the migration streams from South Asia into the Persian Gulf and the concomitant lack of female company have been linked to social frustrations on the part of these migrant workers, frustrations that affect both how local labor markets operate (young men may get so frustrated that they return home) and also overall social and political stability.

In thinking about international labor migration streams, it is important also to recognize that these often follow longstanding cultural and historical links. For instance, in the nineteenth century the migration of Indians to the Caribbean, South Africa, and East Africa was shaped by the fact that both origin and destination regions were part of the British Empire. Likewise, the migration of indentured workers from French India to work on sugar plantations in Réunion or in the French Caribbean (Northrup 2000) reflected the fact that these were all part of the French Empire. Such linkages have continued to shape international labor migration in the postcolonial period – migrants from Angola and Mozambique have found their way to Brazil (Baeninger and Guimarães Peres 2011), as all three countries share a common heritage as Portuguese colonies, while young Algerians have continued to go to France for work, even though Algeria gained its independence from France more than half a century ago.

Finally, in seeking to understand the planetary distribution of labor, it is critical to understand how different economic, social, and political processes – like industrialization or the emergence of a postindustrial society – shape demographic processes, which then affect how many workers there may be in any given society at any given time. In turn, this can influence what kinds of economic activity can occur there and how the labor process is organized; a lack of local labor may force employers in one industry to turn to immigrants, or perhaps to adopt labor-saving technologies, or to increase their wages so as to attract workers from other industries – or even to move somewhere else where labor is more readily available.

CHAPTER THREE

Globalization and Labor

Two ongoing processes are dramatically impacting labor across the planet: globalization and neoliberalization. Although what is meant by these terms is contested, in simple language "globalization" refers to how numerous economic processes and actors now largely ignore national boundaries, while "neoliberalization" refers to how the work patterns and protections that characterized much of the twentieth century in many parts of the world are being swept away and replaced by "precarious work," in which workers have few protections and are increasingly working in part-time and/or short-time jobs. In this chapter, I focus upon how globalization is shaping people's lives (I address issues of neoliberalism and work precarity in Chapter 4). In the first section, I explore some of the ways in which patterns of foreign direct investment (FDI) have created links between working people in different parts of the globe historically and more contemporaneously. In the second section, I look at the phenomenon of global production networks (GPNs), which likewise are linking workers across space. Whereas many have viewed GPNs' form and function as resulting from the actions of trans-national corporations (TNCs), it is important to recognize that workers can also play proactive roles in shaping these entities which are so central to processes of globalization. Finally, I explore the phenomenon of global destruction networks (GDNs), which are in many ways the antithesis of

GPNs – whereas GPNs are involved in putting commodities together, GDNs are networks within which discarded commodities are broken apart so that their constituent elements can be recovered and serve as inputs into new commodities. As with GPNs, so with GDNs: labor plays important roles in shaping how these networks operate and their organizational and geographical structure. This chapter, then, explores how globalization is impacting workers' lives, as well as how working people are playing important roles in shaping how globalization is unfolding.

FDI's Implications for Labor

The past three decades have seen the rise of "globalization talk" (for a history of the word "globalization," see Herod 2009: 47–49). However, what exactly is meant by "globalization" varies, depending upon who is using it. For some, globalization refers to any process in which more "international" things are taking place – the growing availability of "foreign" food in local supermarkets, an increase in the number of foreigners in local labor markets, or the greater ease with which people can travel to different parts of the globe. For them, "globalization" is merely a new name for describing the ever greater delocalization of economic, social, and political life, such that "globalization" and "internationalization" are essentially interchangeable terms for describing the same thing. In such a view, nation-states are still powerful entities and continue to call the shots on economic and political matters. For others, "globalization" has a more specific meaning, namely that the power of nation-states is being undermined by corporations that are increasingly organizing their activities on a planetary basis with little regard for national boundaries. From this perspective, globalization is a zero-sum game – more power to corporations means less power for nation-states. While, for some, globalization is largely an economic process, for others it has political, cultural, and social dimensions. However, despite different interpretations of what globalization is (for more on these debates, see Herod 2009), there is little question that it is transforming the geographical relationships between people located in different parts of the world. This is because the developments in transportation and telecommunications technologies that have facilitated it mean that goods, people, information, and capital can today travel from one part of the planet to another much more quickly than previously. This reduction in the times it takes to travel or communicate between places – a reduction in what geographers call the "relative distances" between them – means that, in many ways, the planet is not as big as it used to be.

The shrinking of the globe is having dramatic consequences. For one thing, it has allowed TNCs increasingly to select their labor force from workers located across the planet and to more readily coordinate geographically separate subsidiaries' actions within globally organized production processes. For another, it has allowed even fairly small firms in one country to more easily service markets thousands of miles away. These developments are having at least three significant impacts upon workers. First, whereas in the nineteenth and twentieth centuries most workers in a country like Britain or the United States were basically in competition only with each other, today millions of workers across the world are in competition with those on the other side of the planet, as many markets for goods and services, as well as many labor markets (for highly educated engineers, for example), have become global. Hence, whereas historically agricultural workers generally produced for local or regional markets because many of the products that they grew were too fragile to survive long-distance transportation, today huge quantities of food travel enormous distances from field to consumer, and producers of many crops are now in competition with those half a world away – Argentine lemons, for example, fill supermarket shelves on the Citrus Coast of Spain as local lemons rot on the ground, while half of Europe's peas are grown and packaged in Kenya, providing competition for European farmers and processors (*New York Times* 2008). Likewise, the Global South's growing industrialization means that whereas a century ago the Global North dominated the production of manufactured goods, and steelworkers or automobile workers in, say, Michigan were primarily in competition with other US workers, today they are also in competition with workers in South Korea, Brazil, and China. Second, the advancements in transportation technologies that allow commodities to be transported from place to place more readily than before mean that manufacturers now feel that they must secure greater flexibility in hiring and firing workers so that they might respond to ever faster changes in market dynamics caused by events on the other side of the world. This need for greater flexibility is helping to drive the growth of precarious work that we will address in Chapter 4. Third, the development of high-speed telecommunications technologies means that even much service work is no longer spatially fixed in place and does not have to be done close to the people being serviced. US or British consumers calling their credit card companies to question a bill, for example, are just as likely to have the call answered in Bangalore as they are in Boston or Leeds, while some cities and states are now subcontracting paperwork processing to overseas vendors – in the early 2000s the City of New York, for instance, outsourced the processing of its quality of life ordinance violations to an office in Ghana (*New York Times* 2002). Similarly, fast-food restaurants like McDonald's have been experimenting with centralized ordering systems for their drive-through operations – the company has established a call center in central California which takes orders from customers in places like Honolulu (Hawaii), Gulfport (Mississippi), and Gillette (Wyoming), and then sends them back to the restaurants via the Internet, to be filled a few yards from where they were placed (*New York Times* 2006). Even some services that were once imagined to have to take place, by necessity, face-to-face have been made mobile – the rise of telemedicine means that a doctor in one part of the globe can now diagnose an illness for a patient elsewhere in real time or that a radiographer in India might process the X-rays of a patient in New York or London.

The transformations described above have emerged in one of three ways. In some cases, independent producers in one part of the globe have simply begun exporting their products or offering their services to other parts of the globe, thereby competing with producers there. In such instances, the lower labor costs and less restrictive environmental regulations in the Global South have frequently put workers in the Global North at a competitive disadvantage, to the point where workplaces in these countries have had to close down. In other cases, TNCs have established subsidiaries in different parts of the globe, either to break into new markets or to take advantage of local labor conditions. In these situations, workers across the planet work under the umbrella of a single entity which is directly managing the entire production process. In yet other cases, entities in one part of the world have hired independent firms in other countries to do work for them (perhaps making components for their products), thereby creating various global production networks. In these cases, workers are brought into relationship with one another but work for legally separate entities. The development of these new relationships between working people across the globe has come amidst a broader transformation in the planet's economic geography. Thus, whereas for much of the nineteenth and twentieth centuries the labor of the Global South was predominantly agricultural, together with some mining and traditional manufacturing (e.g., textiles in precolonial India), and that of the Global North was focused upon industrial manufacturing and, later, service-sector activities, since the 1980s or so there has been a significant reorganization of work globally, as countries like China, Brazil, and India have emerged as manufacturing powerhouses. An important aspect of all of this is that whereas trade is clearly central to connecting the world together, with its attendant consequences for different groups of workers (some may benefit, whereas others may lose out), FDI has now become a more important mechanism for bringing about interconnectedness as, in recent years, its rate of growth has outpaced that of trade (Rainnie et al. 2013).

In order to understand how workers in different parts of the globe have come into greater competition with one another as a new planetary spatial division of labor has emerged in the post-World War II era, it is important to first understand how things used to be. Although several centuries ago both India and China were significant manufacturers whose products were often exported across long distances – Chinese porcelain and Indian textiles were consumed in vast quantities in Europe in the seventeenth and eighteenth centuries, for instance – the impact of European colonialism helped to largely destroy these countries' manufacturing bases. The result was that whereas in 1750 about three-quarters of all manufactured goods were being fabricated in the Global South (many for export), by 1900 most of the world's manufacturing was located in Global North countries and working people in the Global South were primarily either peasant farmers producing agricultural products for local consumption or for export (as in the case of the raw cotton grown in India for shipment to British textile mills) or, in some instances, artisanal producers of manufactured goods sold locally (see Table 3.1). This is not to say, however, that the Global South was not important as a location for profitable trade by colonial trading companies or investment by early TNCs from Europe and the United States. In 1914, for instance, Latin America was the world's largest recipient of FDI, accounting for about one-third of global stock (much of which was invested in mining and plantation operations), while Asia was home to about 21% (Jones 1996: 31). British banks had extensive holdings in Latin America – between 1855 and 1914 the London banking house Rothschild loaned the Brazilian government £173 million, of which about £37 million went to build railroads to link Brazil's coffee plantations with the coast (Shaw 2005: 173) – while British gas companies built street lighting systems for cities like Pará and Pernambuco (Manchester 1964: 325).[1]

In the period prior to the outbreak of World War I, British firms were by far the largest investors globally, accounting for 45% of FDI in 1914 (Jones 2005: 22). This is hardly surprising, given that Britain had the world's largest empire and was its dominant manufacturing and financial power. Indeed, in 1912 one-third of Britain's largest firms had a majority of their assets or production abroad, thereby linking British workers with their counterparts in other parts of the world (Hannah 1996: 148). Other Global North countries' firms were also significantly invested overseas – US firms were responsible for about 14–19% of FDI totals,

Table 3.1 Relative shares (%) of total world manufacturing output for selected countries, 1750–1913

	1750	1800	1830	1860	1880	1900	1913
Europe	23.2	28.1	34.2	53.2	61.3	62.0	56.6
Austria-Hungary	2.9	3.2	3.2	4.2	4.4	4.7	4.4
Belgium	0.3	0.5	0.7	1.4	1.8	1.7	1.8
France	4.0	4.2	5.2	7.9	7.8	6.8	6.1
Germany	2.9	3.5	3.5	4.9	8.5	13.2	14.8
Italy	2.4	2.5	2.3	2.5	2.5	2.5	2.4
Russia	5.0	5.6	5.6	7.0	7.6	8.8	8.2
Spain	1.2	1.5	1.5	1.8	1.8	1.6	1.2
Sweden	0.3	0.3	0.4	0.6	0.8	0.9	1.0
Switzerland	0.1	0.3	0.4	0.7	0.8	1.0	0.9
United Kingdom	1.9	4.3	9.5	19.9	22.9	18.5	13.6
Canada	–	–	0.1	0.3	0.4	0.6	0.9
United States	0.1	0.8	2.4	7.2	14.7	23.6	32.0
Japan	3.8	3.5	2.8	2.6	2.4	2.4	2.7
Global South Overall	73.0	67.7	60.5	36.6	20.9	11.0	7.5
China	32.8	33.3	29.8	19.7	12.5	6.2	3.6
India	24.5	19.7	17.6	8.6	2.8	1.7	1.4
Brazil	–	–	–	0.4	0.3	0.4	0.5
Mexico	–	–	–	0.4	0.3	0.3	0.3

Source: Adapted from Bairoch 1982: 296

Note: All figures are triennial annual averages, except for 1913

German ones controlled 11–14%, Dutch firms 5%, French firms 11–12%, other Western European nations' firms 5%, Japan 0.1%, and the rest of the world 6% (Feis 1930; Woodruff 1966; Jones 1996: 30–31; Bornschier 2000). In many cases, this was invested in these countries' colonies (about 20–30% of all British FDI was in the Empire) and in other parts of the Global South but in many others it was also invested in parts of the Global North – for instance, in the four decades prior to 1914 most French FDI went to countries like Russia and Turkey, for strategic reasons (Berger 2003). Equally, at this time it was the United States that was the world's largest individual receiver of FDI, at about 10% of global totals, after which came Russia, Canada, Argentina, and Brazil (Wilkins 1994).

Whereas European countries dominated FDI in the early twentieth century, by the 1960s the situation had changed significantly. In the two decades after 1945, with the rise of the United States as the world's dominant economy, American firms provided 85% of all new FDI flows worldwide (Jones 2000: 117). They looked particularly to invest overseas because their returns were substantially greater than what they could obtain on domestic manufacturing (Aliber and Click 1993: 86), a situation that arose at least in part because overseas labor was in a weaker position than was US labor. However, two factors – both related to the state of labor in their various economies – conspired to significantly change the nature of FDI flows after 1945 relative to what they had been in the nineteenth and early twentieth centuries. First, as the colonies of the European states gained their independence, many engaged in protectionism and import-substitution industrialization to defend their domestic industries and labor forces from competition from more efficient Global North producers. They were joined by many Latin American countries which had gained independence in the nineteenth century but which worried about growing US economic power. This helped restrict some markets for goods made by US and European workers, with employment consequences for them. Second, many of the Global North's manufacturing industries became more capital-intensive. The combined result of these two shifts was that, whereas prior to World War II much FDI had flowed from the Global North to the Global South as industrialized nations sought both raw materials and markets in the latter, increasingly FDI flowed between Global North nations themselves. Moreover, it was largely focused upon manufacturing and services. Hence, in 1914 about one-quarter of global FDI stock was located in Western Europe and North America. By 1980, however, two-thirds could be found in these two regions (Jones 1996: 31, 48), while Latin America, Asia, and particularly Africa saw their levels of inward FDI drop precipitously – Africa's share of FDI inflows fell from 10% in 1970 to less than 1% in 2000 (UNCTAD 2006: 41). This transformation in the nature and geographical destination of FDI had important implications for labor around the globe and for the relationships between working people in different parts of the world.

In more recent times, the nature of FDI has begun to change again. Although Global North countries hold about two-thirds of global FDI stock (US$16 trillion in 2016, compared to about US$8.4 trillion in the Global South), about 44% of all inflows of FDI (US$765 billion) went to Global South nations in 2015 – US$541 billion went to what the United Nations calls "developing Asia," US$168 billion to Latin America and the Caribbean, and US$54 billion to Africa. Although the United States was the world's largest FDI recipient, at US$379 billion, Hong Kong and the rest of China were the second and third largest (US$175 billion and US$136 billion, respectively). For its part, the United States was the largest FDI provider (at some US$300 billion), while the European Union collectively supplied US$487 billion. However, it is significant that flows of FDI coming from Global South countries have been increasing in recent years (see Figure 3.1) – in 2015, Global South countries provided 25% of global FDI (US$378 billion out of a world total of US$1,474 billion in outflows) compared to only 14% in 1993 (US$34 billion out of a total of US$235 billion). In 2015, for instance, Chinese firms invested US$128 billion worth of FDI overseas and Asia as a whole provided US$332 billion. Firms from Latin America and the Caribbean were responsible for US$33 billion, with nearly half of that coming from Chile alone. Meanwhile, African companies generated only US$11 billion in FDI outflows with nearly US$7 billion of that originating in just South Africa and Nigeria.[2] Nevertheless, some Global South TNCs have started to become big players in the world economy,

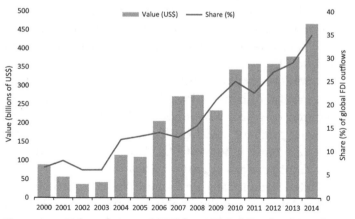

Figure 3.1   Value of outward FDI from Global South economies, 2000–2014

Source: UNCTAD 2015a: 6

including: India's Tata, a company with holdings in steel, beverages, vehicles, chemicals, and hotels that operates in more than 100 countries; Brazil's Vale (mining), Petrobras (petroleum), and Gerdau (steel); China's Shuanghui Group (meat processing) and Sinopec Group (petroleum); South Africa's AngloGold Ashanti and Gold Fields Limited (both mining); Algeria's Sonatrach (hydrocarbons); and Mexico's América Móvil (telecommunications). Although the bulk of FDI from Global South countries remains in the Global South, an important segment is being invested in Global North countries. For instance, in 2014 42% of the US\$1,465 billion worth of FDI coming from the so-called BRICS countries (Brazil, Russia, India, China, and South Africa) went to Global North destinations – the figure for Brazil alone was 83% (UNCTAD 2016a: 12).

Such patterns of FDI have several implications for labor as a resource. First, they mean that workers in disparate parts of the world are being connected together in new and different ways, as TNCs choose where to locate their investments and which workers to hire. Sometimes these decisions are made based upon TNCs' desire to outsource production to cheaper labor – as when US companies relocate the labor-intensive manufacturing of components to Mexico but then import these components back into the United States under the auspices of the North American Free Trade Agreement to use as inputs into commodities being assembled in the United States (see Chapter 6). On other occasions, this FDI is designed to give firms access to new markets overseas and so may be less about finding cheap labor and more about, perhaps, avoiding tariff duties. In both instances, though, through their actions TNCs are creating links between working people in different parts of the world, thereby shaping the kinds of economic and political relationships that different groups of workers have with one another and with local or global employers. Thus, within local labor markets, workers who are employed by local firms may toil under very different sets of conditions than do those who are employed by outside TNCs, with the result that they may be little inclined to collaborate with one another in local labor unions or may have quite different rates of remuneration, which can cause tensions between them. Furthermore, TNCs from different parts of the world are likely to bring their own cultures of work relations to new sites of investment, such that workers employed by subsidiaries of, say, Korean, Brazilian, US, and French TNCs may experience very different work regimes, even within the same community.

Second, it means that the geographical locus of where decisions about investment, technology use, labor relations, remuneration rates, and so forth are made will be very different for workers hired by local firms versus those hired by foreign firms. This has consequences for these two sets of workers – whereas the former may be able to go directly to their local firm's owner and get him/her to change something about their working arrangements, for the latter the ultimate decision-makers may be located very far away and, essentially, uncontactable by local workers. Equally, whereas locally owned firms may be responsive to local pressures and conditions, the subsidiaries owned by TNCs are likely to be less so, as their operational dynamics are often driven more by their place within a TNC's overall organizational structure than by the opinions of local residents or government officials. This has implications for the ability of different groups of workers to shape their working conditions – that is to say, to act as subjects of their own working lives.

Third, TNCs' FDI patterns can often shape labor migration patterns and skill levels, especially when such FDI requires skilled labor that may not be available in destination regions. They can do this in four ways. First, when a TNC establishes a new subsidiary in a region lacking skilled labor, such investment can spur local workers to seek to improve their own skills, perhaps by taking a course at a local technical college. Second, such FDI may prompt skilled workers from the region who have previously left it for greener pastures, either within the same country or abroad, to return home in the hope of being hired. Third, foreign investment may encourage skilled workers from other parts of the country to migrate to where the FDI is being invested, in search of employment. Finally, the lack of skilled workers in the destination location may lead the TNC to bring with it its own skilled workers, such as technicians or managers who have firm-specific technical or organizational skills (Hoxhaj et al. 2015). Overall, then, FDI can be an important mechanism for both encouraging upskilling in the local labor market, as workers perhaps undertake additional education and training, and for encouraging the transfer of technical and managerial knowledge from one part of the globe to another.

So far, we have explored how FDI shapes labor markets and working people's lives as it links them together across the planet. However, whereas FDI represents a situation in which a firm from one part of the globe maintains direct control over the subsidiary it creates abroad, it is also important to recognize that another aspect of globalization is when independent firms act in a globally coordinated manner as part of various global production networks. As is the case with FDI, how GPNs function can have significant implications for labor. This is especially so as the number of workers who are embroiled in them continues to grow – in 1995 some 296 million people (16.4% of those in employment) in the forty countries (representing two-thirds of the global labor force) for whom data were available were employed in global supply chains; by 2013 that figure had grown to 453 million people (20.6%) (ILO 2015: 132–3).

GPNs and Labor as Object and Subject

Simply put, GPNs are assemblages of firms involved in the production of commodities. Sometimes also referred to as global commodity chains or global value chains, they are typically fairly hierarchical entities and involve independent manufacturers and support companies located in different regions of the globe all working as part of a single network to produce a commodity – a car, a computer, a shirt, and so forth. Sometimes these involve producers in the same economic sectors (as when computer chip manufacturers produce components for companies like Dell or Hewlett-Packard) and sometimes they involve producers in quite different sectors (as when plantation workers in West Africa grow cocoa that will be turned into chocolate by European factory workers for sale in North America). Significantly, although coordinated as singular wholes, specific parts of the GPN may be located in different areas around the world so as to take advantage of particular conditions there – close access to certain raw materials, cheap labor, lax environmental regulations, etc. Hence, particularly hazardous parts of the production process may occur in countries with few environmental regulations, labor-intensive segments may take place in countries with cheap labor, and the technically complex aspects of the process may occur in Global North countries that have ample access to precision machinery requiring few workers. The result is that components may be shipped significant distances before they find their way into a final product.

Typically, it is the firms making up any given GPN – and especially those at its organizational apex – that are seen to shape the network's structure and thus to determine how workers in different parts of the globe are linked together through various "strategic couplings." These couplings involve processes of corporate *layering* (in which TNCs make successive rounds of investment in particular regions), *conversion* (in which TNCs change the nature of their relationship with a region, perhaps through changing their production process in particular plants), and *recombination* (in which a region's assets are linked in new ways to additional waves of corporate investment, which itself can involve a TNC abandoning another community – "decoupling" from it – and thereby transforming the geographical configuration of the overall GPN; see Yeung 2009; Rainnie et al. 2013). Clearly, each of these aspects of strategic coupling and decoupling can have significant implications for workers incorporated within the GPN (or recently disconnected from it). New investment in a region ("layering") may lead to more workers being hired but may also cause some to be laid off if the new investment happens to be that of labor-saving technology. Likewise, "conversion" may result in workers with skills that were appropriate for one production process being fired while workers with different ones may be hired to manufacture new products or to manufacture the same product but using different technologies – perhaps the firm has upgraded its machinery and needs workers with greater computer skills to operate it. Equally, by changing how various firms within the GPN interact with one another and with local institutions, recombination may have significant impacts upon these firms' labor forces and how they are connected across space, and also how they interact with one another within the production process in their individual workplaces.

Although the TNCs that sit atop GPNs, then, are often viewed as coordinating how they function and therefore how their constituent firms – and the workers employed therein – interact with one another, in fact this is not always the case. First, whereas they are often viewed as relatively powerless, in fact small firms at the "bottom" of the GPN (perhaps fourth- or fifth-tier subcontractors) can sometimes exert significant influence over how the network operates, especially in cases where GPNs operate according to just-in-time (JIT) methods of production and inventory control.[3] Any disruption of the supply of components from such small firms can have significant ripple effects throughout the GPN. By way of example, the 2011 Japanese tsunami impacted GPNs far beyond Japan because numerous lower-tier suppliers were shuttered – Toyota halted production at several US assembly plants because it could not get parts from Japan, while Ford told its US dealerships to suspend taking orders for vehicles painted in "tuxedo black" and to limit orders in three shades of red because one of the components in these paints could not be secured from the only plant making it, which had been shut down by the tsunami (Herod 2011).

Second, and relatedly, the workers who are embroiled in a GPN can also have significant impacts on how it functions. For those who work at the apex of the GPN this is easily understandable. For instance, they may go on strike at the final stage of a product's assembly, which will likely have significant implications further along the network, as it is probable that components used in a product's final assembly will not be needed during the strike's duration. This will affect both workers at these components' manufacturing plants (they may be laid off) and their communities (which will see less money cycle through them because laid-off workers are unlikely to be purchasing as many goods and services as they had been). However, workers at the bottom of the GPN pyramid can also dramatically shape its organizational form and how it functions through their actions. In 1998, for instance, when just 9,200 workers walked off the job at two plants supplying General Motors with parts, GM's production chain was so hampered that it ultimately had to temporarily lay off more than 193,000 employees and close 27 of its 29 North American assembly plants and 117 components' supplying factories in Canada, Mexico, the United States, and even as far away as Singapore. Moreover, the domino effect of the strikes at these two plants ended up costing GM about 500,000 vehicles in lost production, making it one of the most expensive strikes in US history (Herod 2000). What this shows is that workers are not just objects of manipulation by TNCs and other firms involved in GPNs; they are also subjects of their own actions, as they can proactively impact how the GPN functions and is structured economically and geographically.

Waste, Global Destruction Networks, and Labor

The discussion above explored how FDI and GPNs have linked together working people across the world. It also illustrated how workers both have their lives shaped by processes associated with globalization but can also play significant roles in shaping them. However, in thinking about how commodities are put together in GPNs, it is also important to think about what happens to them at the end of their useful lives, once they are discarded. This is because the production of waste is a central aspect of how capitalism operates – consumers are frequently encouraged to upgrade to new versions of a product as a way for firms to sell ever greater quantities of goods (the iPhone is iconic in this regard) – while waste's movement across the planet has significant effects upon different groups of workers. It also has momentous impacts upon the environment and for the global political economy, with Global North waste often exported to the Global South for disposal.

While the manufacture of commodities in GPNs sees these commodities move from the hands of workers in one place to those in another, similar webs of connections between workers in different places can be involved in the taking apart of commodities at the end of their lives. These webs – what have been called global destruction networks (GDNs); (Herod et al. 2014) – likewise link workers across space as they disassemble unwanted commodities to recover any valuable components for reuse in new commodities. Arguably, of all types of waste, it is e-waste that has seized the popular imagination in this regard, especially as some electronics firms have sought to gain goodwill by publicizing their efforts to recycle their products – Apple, for instance, has made much of its e-waste take-back program, which allowed it to harvest in 2015 some 23,101,000 pounds of steel, 39,672 pounds of nickel, 13,422,360 pounds of plastics, 44,080 pounds of lead, 11,945,680 pounds of glass, 130,036 pounds of zinc, 4,518,200 pounds of aluminum, 4,408 pounds of tin, 2,953,360 pounds of copper, 6,612 pounds of silver, 189,544 pounds of cobalt, and 2,204 pounds of gold, with this later valued at about US$40 million (Apple Inc. 2016). Many other products are also taken apart within GDNs – ships, for example, are typically sailed to places like India and China to be cut up and various components recovered from their hulks. Some of these are then reused pretty much as is. Hence, the engines of ships broken up in India are frequently sold to textile factories, and their boilers end up in rice mills (Hossain and Islam 2006). Equally, tables and chairs may be sold in local markets for people to purchase for use in their homes. Other parts, like steel, may be smelted down and turned into ingots for use in new products like construction girders or door panels in the automobile industry.

GDNs, then, are important globalization agents affecting labor in different parts of the world and are central in the process of moving used commodities and the still-valuable components they contain from one place to another – say from wealthy consumers in the Global North to informal workshops in India or Ghana, where they are dismantled under typically fairly horrendous conditions (small children are often exposed to noxious fumes from melted plastic, for instance, as they burn off the protective coatings from wires so that they can get to the copper within them). Equally, GDNs have important implications for the communities in which such waste ends up. For instance, whereas some of the precious metals, plastics, glass, and so forth recovered from discarded commodities may have value and are sent to other parts of the world for use as inputs into new products, the remainder often end up simply being buried in landfills close to where the discarded products were disassembled. This may provide employment for local workers, but it can also expose them and their families to environmental hazards, including toxins leaching into groundwater from landfills that have not been properly lined. This has effects on workers' abilities to reproduce themselves daily and generationally. GDNs, though, also have implications for working people in resource-producing communities, for the more that metals and other components are recovered from old commodities, the less do fresh quantities of such components need to be dug out of the ground or harvested, which may encourage unemployment in these communities – one study has estimated that recycling the five to six million tonnes of metal discarded in Australia each year could provide 60–70% of the country's annual needs (Corder and Golev 2014), with potentially significant consequences for iron ore miners (see Chapter 5 for more on Australian mining).

In terms of how any given GDN is structured geographically, labor's characteristics in different parts of the globe – its wage rates, the environmental regulations under which it toils, its skill levels, and so forth – can certainly shape where those who are orchestrating the GDN choose to locate certain elements of it (i.e., they view the labor involved as object). At the same time, however, much as with GPNs, GDNs' structures can also be substantially shaped by the actions of those workers embroiled within them. Workers as subjects, in other words, can shape how GDNs are constituted. In the case of e-waste, workers can influence a GDN's form through the types of labor they are willing to do. In the Global North, although some components like Cathode Ray Tubes from TVs are taken apart by hand, because e-waste often contains fairly toxic substances like mercury and lead its dismantling is generally done using large-scale machinery within the formal economy. For example, Apple has designed a line of robots – the company calls them "Liam" – that are capable of taking apart an iPhone every eleven seconds, sorting out its various quality components for recycling. This is a much quicker process than dismantling them by hand, allowing Apple to recycle 1.2 million iPhones annually. Not only does the capital-intensive nature of the labor process mean that relatively few workers are involved, but it also affects the types of materials that are recovered. Hence, after some initial sorting to remove easily collected components like batteries, or to see if a monitor from a discarded computer might still be usable, with such parts often either resold locally or online, the use of machinery in the Global North means that the remaining parts are usually broken up into fist-sized pieces that are then melted down to retrieve various metals, plastics, and the like.

In the Global South, by way of contrast, the disassembly process is typically more labor-intensive and conducted within the informal economy. Hence, once e-waste is delivered to the communities where it will be processed – places like Bangalore, India, and Agbogbloshie, a suburb of Accra, Ghana that is sometimes called "the largest e-waste dump in Africa" – local people (often children) take it apart using hand tools and sort it so that reusable working components like circuit boards and hard drives can be extracted and sold directly to local technicians, who then use them to make non-brand-name goods or to fix broken electronics. Some of these reusable parts are even exported abroad – printed circuit boards recovered in Ghana have been shipped to China for reuse. The remaining items are then usually processed to extract the valuable elements, frequently through burning or using strong acids. What is significant in all this, then, is that the type of labor involved shapes dramatically the kinds of components recovered and thus the dynamics of the GDN and what can be sent on as inputs into GPNs making new commodities. Thus, in the Global North, the use of complicated and expensive equipment like large hammermill machines and industrial-sized magnets and eddy-current separators to divide ferrous and non-ferrous metals means that rates of recovery of constituent elements are usually higher than in the Global South. On the other hand, the fact that labor costs are much lower in the Global South and that there is less access to expensive equipment means that more of the work of breaking up e-waste and retrieving its components is done by hand. Consequently, whereas in the Global North much of the e-waste is simply melted down to recover raw materials, in the Global South a higher proportion of parts from computers and other electronics are more frequently recovered whole and repaired and reused directly in new products.

With regard to shipbreaking, the labor process can also have significant impacts upon how the GDN is structured. For instance, in places like India and Bangladesh, ships are typically beached at high tide and then an army of local workers who are informally employed go to work breaking them up. Typically, they use tools that are no more complicated than screwdrivers, sledgehammers, and perhaps oxy-acetylene cutting torches. Such work is extremely hazardous, both because workers are exposed to toxins and because of the potential for accidents – workers are frequently maimed or killed by falling metal or sharp edges when the ships' hulls are cut apart. By way of contrast, in China ships are usually dismantled in docks using large-scale machinery. The primary reason for this is that Chinese workers are not willing to tolerate the poor working conditions and low remuneration levels that Indian and Bangladeshi workers will accept, with the result that breaking firms have been forced to modernize their operations with labor-saving technologies. The latter's techniques are also seen as being more environmentally friendly, and this can give them an advantage with some consumers who want to purchase "greener" products.

Finally, in considering the relationships between GDNs and GPNs, it is important to understand how the labor process in one can shape that in the other. As indicated above, the dissimilar ways in which workers toil in the Global North and the Global South result in different types and amounts of components being secured from old products, which can affect how GPNs that are reliant upon recycled materials are structured. Furthermore, the growing emphasis upon recycling as a way to reduce using finite resources and to save on energy and other costs means that the design of many types of products is also changing. This is having impacts upon how work is organized within both GPNs and GDNs. In particular, increasing numbers of manufacturers are adopting production models called "designing for recycling" or "designing for disassembly," wherein products are designed specifically so that their disassembly is made much easier. This is shaping how they are then taken apart at the end of their useful life, with weighty consequences for how workers in various GDNs labor.

Summary

Globalization is having momentous impacts upon workers. TNCs are increasingly playing workers in different parts of the world against one another in a "race to the bottom," wherein people compete to work for less than their competitors or agree to give up workplace and environmental protections for fear that failing to do so will result in them having no jobs. TNCs, then, are reshaping the planet's economic geography and where work is to be found by using geography creatively, pitting labor in one place against that in another as part of their strategy for securing profits. The massive amounts of FDI being unevenly invested across the planet are reshaping the life chances of hundreds of millions of people, as some see FDI and jobs come to their communities, while others see capital disappear from theirs and suffer the effects of this (such as unemployment). At the same time, FDI is shaping patterns of international and intranational labor migration, while GPNs and GDNs are linking workers across the world in novel ways. However, it is also important to recognize how working people themselves are affecting processes of globalization through, for instance, shaping how various GPNs and GDNs are structured geographically and so function organizationally. In such regards, working people continue to be both objects and subjects in the remaking of the planet's labor markets.

CHAPTER FOUR

Neoliberalism and Working Precariously

If globalization is one process impacting labor across the planet, the rise of precarious work is another. The concepts of "precarious work" and "work precarity" have become quite ubiquitous in recent years and have led at least one commentator to term the workers who toil under such conditions "the Precariat" (Standing 2011), by which he means they are a class of individuals only tangentially involved in the labor market and, when they are involved, the conditions under which they labor and their levels of remuneration are quite poor. According to the International Labour Organization (ILO 2015), today only about 25% of workers worldwide have any kind of stable employment relationship. The remainder are entwined in so-called "contingent" relationships, working on temporary or short-term contracts, in informal jobs, as "independent contractors," or in unpaid family jobs. For sure, not all contingent work is necessarily precarious – highly skilled lawyers or engineers may only work periodically out of choice, earning a good living when they do so. However, many contingent workers do face growing problems of precarity, as securing sufficient income on which to live and raise their families seems ever more difficult. Although the prevalence of precarious work varies geographically, in virtually every corner of the globe one thing is fairly unvarying: it is growing in extent. In the United States, for instance, whereas about 17% of workers were contingent in 1989 and 30% in 2005, today the figure is over 40% and likely to continue rising (Intuit 2015; US Government Accountability Office 2006 and 2015). In the UK, more than one-fifth of the labor force is contingent, while in Australia a similar proportion is on some form of casual contract. In France, permanent jobs account for just 16% of new contracts, compared to a quarter in 2000. In Spain, nearly 70% of young workers labor under temporary contracts. Overall, within the European Union just over half of all young workers are in temporary jobs, the highest figure on record (*Financial Times* 2015). Work that is potentially precarious is also becoming increasingly common in other parts of the world, such as in Asia – in Japan about 40% of the labor force works contingently and in South Korea about one-third does. For its part, in China almost all the jobs created during the past three decades have been fixed term (i.e., temporary) (ILO 2012), while fewer than 10% of the millions of migrant workers involved in producing that country's economic growth in recent years have a signed labor contract, let alone secure employment (Chan 2009). In fact, in many countries around the world anywhere between 30% and 45% of the working-age population is unemployed, inactive, or working only part time – in just the United States, the United Kingdom, Germany, Japan, India, Brazil, and China alone, some 850 million people (about 25% of the total global labor force) fall into this category (McKinsey and Company 2015).

The expansion of precarious work is occurring within the context of the entrenching in many economies of an increasingly bifurcated labor market. On the one hand, there are workers who enjoy long-term work, who are generally reasonably well remunerated, and who commonly work in comparatively good conditions. These workers serve as a core labor force. On the other hand, there are many others who are typically employed as temporary, part-time, or even full-time workers on low wages, with little security, and under poor conditions. These serve as a peripheral workforce, often servicing the core group of workers. Although such workers are overwhelmingly those who have historically often been most marginalized in the labor market (women and racial/ethnic minorities; see Tables 4.1 and 4.2), in recent years the number of formerly privileged workers (males and members of the racial/ethnic majority in a particular area) who are caught up in its grip has also been growing as precarity spreads across ever larger sectors of various economies, thanks to the removal of many of the protections that workers – particularly those in the Global North, though not exclusively – enjoyed for much of the twentieth century and to the growing desire for labor flexibility on the part of many firms.

This chapter explores the rise and extent of precarious work. First, I detail what is meant by neoliberalism and

Table 4.1 Percentage employed full time for an employer, among the entire adult population, ranked by women's deficit

	Men	Women	Deficit
South Asia	33	9	−24
Middle East and North Africa	31	8	−23
Latin America and Caribbean	36	19	−17
Southeast Asia	29	13	−16
European Union (EU)	43	28	−15
East Asia	35	21	−14
Non-EU Europe	34	22	−12
Former Soviet Union	48	37	−11
North America	48	38	−10
Sub-Saharan Africa	14	8	−6
Globe	34	18	−16

Source: Gallup 2014b; based upon surveys in 163 countries and areas in 2013

Table 4.2 Percentage employed full time for an employer, among those in the labor force, ranked by women's deficit

	Men	Women	Deficit
Middle East and North Africa	44	29	−15
Southeast Asia	35	23	−12
European Union (EU)	64	52	−12
East Asia	45	33	−12
Latin America and Caribbean	48	38	−10
South Asia	47	38	−9
Sub-Saharan Africa	19	12	−7
Non-EU Europe	52	49	−3
North America	63	60	−3
Former Soviet Union	63	62	−1
Globe	46	36	−10

Source: Gallup 2014b; based upon surveys in 163 countries and areas in 2013

some issues when it comes to defining precarity. In so doing, I also make connections between the rise of precarious work and the kinds of challenges faced by firms as a result of globalization. I then examine different types of precarious work, how they have been growing in recent years, and what they mean for labor.

Neoliberalism and Precarious Work

Although the term "neoliberalism" (and therefore "neoliberalization") was first coined in the 1930s, much like "globalization" it has become much more frequently used by academics, activists, and others in the past three decades or so. While different people mean different things by it, in general "neoliberalism" refers to a state of affairs in which labor market and workplace protections for workers have been reduced and/or removed so that labor markets can operate more "freely," in which the guiding economic philosophy is one whereby "the market" decides what is best for society, and in which the role of government has increasingly shifted from that of ensuring that public goods (like education and healthcare) are distributed equitably throughout society to ensuring that market competitiveness is fostered through providing economic incentives to firms and slashing regulations and tax rates. Despite the fact that it includes the term "liberal," then, neoliberalism is actually a conservative economic philosophy, with "liberal" – from the Latin *liber*, meaning free – referring to "free markets" and "free trade." Frequently, "globalization" and "neoliberalization" are used somewhat interchangeably, though this is an incorrect linking of the two, as it is possible to have non-neoliberal globalization – international worker solidarity could be considered a form of globalization, though it would not be a form of neoliberalization. Moreover, whereas many see neoliberalization as emerging because the state is withdrawing from the role of regulating numerous aspects of the economy, in fact the neoliberal policies that have encouraged precarity – policies like implementing free trade agreements, privatizing various government services, and "deregulating" the economy – have only been possible because governments have opted to pursue them.[1]

As an economic philosophy, neoliberalism was arguably first adopted in wholesale fashion in Chile after the 1973 military coup that brought to power dictator Augusto Pinochet. The economists hired by the new regime had largely trained at the University of Chicago, where they worked under professor Milton Friedman and were hence known as the "Chicago Boys." Viewing Chile largely as a vast, real-world laboratory in which to try out their "supply-side" theories, the policies that they and their government sponsors adopted included privatizing and "deregulating" much of the economy and significantly reducing and/ or eliminating many trade barriers. They also "reformed" labor relations by undermining the trade unions (with many unionists ending up being murdered by the regime or going into exile) and by introducing market competition into formerly government-provided social services, like healthcare and education. Whereas the richest Chileans saw their incomes rise, for the country's working classes living standards plummeted dramatically under such policies. Similar policies were adopted by other governments in both the Global North and the Global South in the following decades. In the United Kingdom in 1979, for instance, under the slogan "there is no alternative," the Conservative government under Margaret Thatcher began pursuing a neoliberal policy of tax cuts, labor market deregulation and attacks on organized labor, and the privatization and/ or cutting of government services. In the United States, the Reagan Administration followed similar policies. Even some putatively socialist governments have pursued policies influenced by what would come to be called neoliberal thinking, such as the governments of François Mitterrand in France and, in China, of Deng Xiaoping, who was a fan of Friedman and who introduced what would euphemistically come to be called "socialism with Chinese characteristics." With the collapse of Soviet communism in the late 1980s, that country's former Eastern Europe satellite states were subjected to various neoliberal "shock therapies" designed to jump-start their languishing economies. More recently, President George W. Bush used post-invasion Iraq as a place in which to experiment with neoliberal policies (Looney 2003).

Within this context, precarious work has mushroomed. In considering what this amounts to, there are two important things to ponder: defining what is meant by the term and understanding why it appears to be becoming more common. In terms of the first of these, specifying what is meant by precarious work is sometimes harder than it might at first seem. Typically, it is defined in contrast to "standard work," although that merely raises the question of what is meant by standard work. Defining precarious work has also been complicated by the fact that other terms, such as contingent work, atypical work, and non-standard work, are often used to describe the kinds of conditions under which many working people labor, and this has been compounded by the fact that these terms' meanings have shifted over time. For instance, the phrase "contingent work" was first used in print by economist Audrey Freedman (1988) when referring to a management technique in which labor is hired when there is an immediate need for its services and then laid off when that need has passed. Since then, it has expanded to include a much wider range of work practices, such as part-time work, temporary employment, self-employment, and home-based work. However, all these work types cover a variety of experiences that are quite different from one another. Hence, although home-based work might be considered "contingent," as a category it can include both self-employed lawyers working from home in the suburbs who set their own hours and fees and work pretty much when they choose, as well as the activities engaged in by an immigrant woman in a poor urban neighborhood who is sewing buttons on shirts and being paid by the piece. Nevertheless, in general, precarious work is understood to mean work which is undertaken involuntarily for low pay, with few if any benefits, and frequently under dangerous circumstances wherein workers have little regulatory protection from either the state or a union contract. Perhaps its single most important characteristic, though, is its very lack of certainty and security.

Thus, the ILO (2012: 27) defines precarious work as allowing employers to involuntarily

> shift risks and responsibilities on to workers. It is work performed in the formal and informal economy and is characterized by variable levels and degrees of objective (legal status) and subjective (feeling) characteristics of uncertainty and insecurity. Although a precarious job can have many faces, it is usually defined by uncertainty as to the duration of employment, multiple possible employers or a disguised or ambiguous employment relationship, a lack of access to social protection and benefits usually associated with employment, low pay, and substantial legal and practical obstacles to joining a trade union and bargaining collectively.

In seeking to understand why precarious work is becoming more common, some of the explanation undoubtedly lies in the choices that workers themselves are actively making (i.e., labor as subject). Hence, in many Global North countries, longer life expectancies mean that many seek out part-time work after they have retired just as a means of keeping busy. Others prefer such work because it may fit more easily with their lifestyles or because they want a better work–life balance than that offered by full-time employment. For still others, they may prefer not to be tied to a particular employer for years but to change employers and fields of occupation periodically for new challenges or even to branch out on their own in self-employment. There are also some people who may choose to work part time because they are able to make a living through, say, successfully investing in the stock market using various online trading media in ways that were not possible a few decades ago. It is also the case that part-time or temporary work or unpaid internships and voluntary work can sometimes be a stepping stone to full-time and/or paid employment, such that many workers (like young people just entering the labor market) may be willing to do it to get a foot in the door. However, the majority of precarious workers are engaged in such work not through choice but because they have no alternative and this is the only kind of work offered them by the firms seeking to hire them (labor as object). Indeed, innumerable studies have shown that millions of workers who are stuck in precarious employment would prefer more secure, full-time employment.

Several factors help explain involuntary precarious work's growth. For some employers, the ability to hire workers on a short-term basis allows them to accommodate fluctuations in the amount of work they have – hiring at busy times and firing when the need has passed – or to fill in for workers who are absent due to illness or because they are striking. It also allows them to evaluate workers before considering them for permanent employment. In countries like the United States, where part-time employees are not entitled to some of the benefits that full-time workers receive (such as health insurance or pensions), it also provides a means to cut the costs of hiring workers – hiring two workers to each work twenty hours a week can be cheaper in the long run than hiring one worker for forty hours, even if the hourly wage is the same in both cases. Firms can also threaten full-time workers with replacement by part-timers should they get too demanding. Hiring contingent workers can correspondingly give firms access to specialized labor that they may only need for a short period of time or for a particular project (say, upgrading a computer system). Much of this flexibility has been facilitated by the rise of large firms that can supply temporary workers – companies like the ManpowerGroup and Kelly Services in the United States, BlueArrow in the United Kingdom, SEEK, Chandler Macleod, and Hays Specialist Recruitment in Australia,

Recruit Holdings in Japan, and the French-Swiss Adecco Group. Although in the 1940s and 1950s such "temp agencies" typically provided low-skilled office workers to cover for, say, a secretary who was on vacation, over time the types of workers they provide has expanded significantly – many factory workers are now hired through such agencies (who are their legal employers) and the temps hired frequently have significant skill sets (high-level computer skills, legal or management expertise, and so forth). The result is that, whereas in the past temp agencies were frequently a last resort for firms looking for labor at short notice, over time they have become permanent labor-market intermediaries supplying workers on a long-term basis. The growing use of these workers as a central and persistent part of firms' labor forces means that many are actually now "permatemps," toiling in the same workplace for years on end. However, because they work for the temp agency rather than for the firm which owns the workplace, they typically have few job protections and often are paid less than the workers who are directly hired by the firm, even when doing the same work.

The growth of precarious work, then, has been driven by the neoliberalization of labor markets and economies more broadly. But it has also occurred within the context of the challenges that globalization has presented numerous employers. As outlined in Chapter 3, relative distances between many parts of the planet are being diminished by advances in telecommunications and transportation technologies, and this is dramatically reducing the time in which firms have to react to market changes caused by things like news of a strike or a political assassination on the other side of the world – a fact readily seen by observing the changing prices of shares as they scroll across the bottom of your TV screen whenever a significant event occurs somewhere. Geographer David Harvey (1989) has referred to this phenomenon as "time–space compression," while sociologist Ben Agger (1989) has suggested that these developments mean that we are now in an era of "fast capitalism." Consequently, firms have been encouraging precarity among their labor forces by seeking ever greater labor flexibility so as to be able to respond to these changes quickly. Such flexibility takes one of two forms (though many firms have adopted both simultaneously). In some cases, firms have sought out *numerical flexibility* (the so-called "American model" of flexibility). The goal here is for firms to be able to more readily expand their labor force when needed and to reduce its size when not. Sometimes this has involved them turning to the external labor market, either hiring more workers for short periods of time and then dismissing them when no longer needed, or subcontracting out work. At other times, it has involved reconfiguring their internal labor forces' activities, by forcing employees to work overtime when needed and/ or cutting back their hours on other occasions, or through greater reliance upon "flextime" in which workers vary their starting and stopping times, sometimes on the basis of their own needs or desires but frequently according to those of their employers. In other cases, employers have sought out greater *functional flexibility*, in which firms secure the ability to more easily redirect their employees within the work process, such as by having a worker do one task one hour and a different task the next. This so-called "European model" of flexibility looks to reduce the numbers of workers a firm might have to hire, as having one worker who can do multiple tasks eliminates having to hire several, each of whom can only do a single one. However, this "teamwork" model of employment relations is seen by many as incompatible with older forms of labor unionism, wherein unions typically insisted upon rigidly delineated task responsibilities as a way to protect their members' jobs, which they "owned" by dint of the fact that they were the only ones allowed to do them. In turn, this has led many employers to attack unions' prerogatives in the workplace because the unions are seen as inhibiting flexibility and the initiation of teamwork schemes. This has resulted in the elimination of many workplace protections and thus greater insecurity for long-term workers.

Significantly, precarity affects workers in virtually every segment of the economy today, both those in the so-called "Old Economy" (activities such as mining and manufacturing, upon which industrialization was based) and those in the "New Economy" ("postindustrial" jobs like "knowledge working," which were held out in the 1990s as being the future of work under capitalism) (see Chapters 5 and 6 for more on work in the Old and New Economies). Thus, in the Old Economy automation has reduced the need for labor, while attacks on labor unions have dramatically undermined workers' protections and economic bargaining power. This has been combined with efforts to privatize many sectors of the economy by selling off formerly nationalized industries like coal (in places as diverse as the United Kingdom, Russia, India), railways (the United Kingdom, Pakistan, India, Japan, Germany, the United States, Australia), airlines (many countries across Africa, as well as in Europe), utilities like water, electricity, and gas, and many others. Even in putatively socialist/communist states, many state-owned enterprises have been sold to private investors, with significant implications for workers, many of whom have lost their jobs while those who remain have often seen working conditions deteriorate. In China, for instance, the government closed thousands of state-owned firms in the 1990s and in 2015 announced that it would diversify ownership of thousands more by making it easier for private capital to own stakes in them. It is also having the day-to-day running of many workplaces done not by the state-appointed bureaucrats who have run them for years and whose goal has been to ensure that government production targets were met, regardless of profitability, but instead by professional managers who are supposed to make them profitable – goals that will put greater pressure on workers and no doubt increase their sense of insecurity. Even in countries where labor costs have historically been fairly low, there are growing pressures on workers. Hence, the ILO (2016) estimates that in the next two decades some 137 million workers (56% of the labor force) in Cambodia, Indonesia, the Philippines, Thailand, and Vietnam will lose their jobs due to automation. Similarly, precarity is evident in the New Economy of knowledge workers and those engaged in "gig employment" in the "on-demand marketplace" – for example, those who drive for Uber, or who deliver groceries that have been ordered through phone apps like Instacart, or who work as on-demand couriers for firms like Postmates (which uses mobile phones to receive orders and dispatch deliveries), or who are registered with Amazon's Mechanical Turk (whose website claims to "give businesses and developers access to an on-demand, scalable workforce" and workers the ability to "select from thousands of tasks and [to] work whenever it's convenient"). Work in the New Economy, then, is also becoming highly stratified, with a handful of people at the top being responsible for various creative activities (originating knowledge), while a significant number at the bottom simply manage it, frequently in insecure circumstances (as with the stereotypically ubiquitous data processor working in a cubicle along with dozens of other such workers and subject to the

Taylorist work discipline reminiscent of nineteenth-century factory work).[2]

It is important to consider the precarious lives experienced by growing numbers of workers not only in economic terms but also in broader environmental and biological terms. Thus, with growing attacks by the state and employers on the health and safety rules negotiated by labor unions over time, together with greater desperation on the part of many working people who will take whatever job they can, substantial numbers of employees now find themselves exposed to toxins and other dangers in the workplace from which they were previously protected, exposures that are having significant impacts upon their bodies over both the short and the long term – everything from greater exposure to accidents at work to cancers caused by contact with workplace chemicals. Equally, the stress of precarity can cause hypertension and other types of disease that may cut short an individual's working life, with consequences not only for the individual but also for family members who may have to care for him/her once he/she becomes too sick to work. There are also, though, more far-reaching aspects of the precarity that millions of working people are facing. In particular, myriad peasant farmers, herders, and others across the Global South are confronting significant challenges from global climate change. This is especially so in Africa, where climactic changes are among the planet's worst and where poverty – largely a legacy of colonialism – means that governments, other institutions, and individuals are not well positioned to respond. Hence, it is estimated that some African countries – particularly those of the Sahel, but elsewhere too – may see agricultural yields drop by 50% in the next few years, a situation that will not only make sustaining human life much more difficult but is also likely to lead to significant political instability, thereby exacerbating the problem. This is also going to shape patterns of FDI, both directly (as certain regions become increasingly off limits to investors due to a lack of resources like water and labor capable of work, thanks to its weakened state) and indirectly (as investors seek to avoid areas of political instability). In other areas, like the highlands of Ethiopia and Kenya, changing rainfall patterns are likely to encourage the spread of malaria. In Asia, meanwhile, rising sea levels and a warming atmosphere will worsen monsoonal flooding, threatening millions of people's abilities to survive – parts of some Pacific islands are already being drowned by such developments, with serious consequences for their inhabitants. Significantly, many of the people affected by these changes are likely to become "climate refugees" and thus, due to the fraught situations within which they find themselves, ripe for economic exploitation wherever it is that they end up living. Ecological degradation, then, is intimately connected to social and economic insecurity and thus to the precarity faced by hundreds of millions of working people.

Forms of Precarity and Their Present Dynamics

According to the ILO (2015: 13), only about half of global employment is made up of waged or salaried jobs, though this varies significantly by region – in the Global South employment in the informal economy, often in unpaid family work, dominates and relatively few people are on a payroll (see Table 4.3). Significantly, these types of informal employment relations also dominated in the Global North up until the late eighteenth century. With the spread of industrialization, though, employers became increasingly reliant upon having a regularly available labor force

Table 4.3 Global high and low payroll to population (P2P) rates, measured as a percentage of the population aged 15 and older, 2013

	Highest P2P rate		Lowest P2P rate
United Arab Emirates	59	Burkina Faso	5
Iceland	54	Haiti	6
Bahrain	53	Malawi	6
Sweden	53	Niger	6
Russia	51	Ethiopia	7
Kuwait	49	Sierra Leone	7
Belarus	47	Guinea	8
Israel	44	Liberia	8
Latvia	44	Mali	8
United States	43	Chad	9
		Nepal	9
		Tanzania	9
Globe	*26*		

Source: Gallup 2014a; based upon surveys in 136 countries

on hand. This required more formal employment arrangements with workers, such that for much of the past two centuries workers in the Global North have typically worked in relatively stable, full-time jobs – the so-called "standard employment" model. In both the Global North and the Global South, however, many of the jobs that used to be done under standard employment conditions are now being replaced by work forms generally seen as symptomatic of precarious work. In fact, the ILO estimates that fewer than 40% of waged and salaried workers globally are employed on a full-time, permanent basis. The rest are employed on temporary or short-term contracts in informal jobs that frequently do not have any contract, in unpaid family jobs, or as contract workers. Women are disproportionately represented in such employment. Moreover, the ILO has found that in the majority of countries in which reliable data are available, the growth in part-time jobs has been outpacing that of full-time jobs in recent years.

In the case of temporary work, the ILO (2012: 31; 2015: 30) estimates that the proportion of the global labor force working under such conditions grew from 9% in 1985 to 13% in 2015. However, this varied considerably across the planet (see Figure 4.1). In high-income Global North countries, where more than three-quarters of paid workers are on a permanent contract (working either full time or part time), about 9.3% of workers are temporary contract workers and the remaining 14% work without a contract. In the thirteen low-income countries (covering 49% of total employment in low-income countries) for which the ILO has reliable data, on the other hand, only about 5.7% of workers are on a permanent contract, 7.5% are temporary contract workers, and the remainder have some other kind of arrangement, typically working for themselves informally or as unpaid family workers. Although temporary work is less common in rich Global North countries and generally more common in poorer ones, there is nevertheless significant variation across the wealthy countries – within the European Union, for instance, Poland has a much greater proportion of its labor force employed in temporary work than do the UK and Belgium. While many temporary workers are hired directly by the firms owning the workplaces within which they work, significant numbers are hired via employment agencies – in 2013 some 60.9 million workers globally gained access to employment this way, of whom 40.2 million worked directly as an agency worker (11 million in the United States, 10.8 million in China, 8.7 million in Europe, and 2.4 million in Japan) (CIETT 2015: 10). Many of these

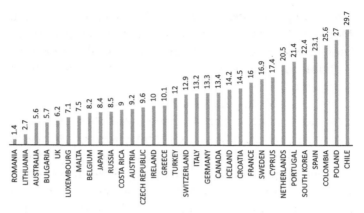

Figure 4.1   Temporary employment as a % of dependent employment, 2013

Source: OECD 2016b

are permatemps, working for years on "temporary assignments" in the same or different workplaces.

Part-time work is also growing globally, especially in places where it has historically been less common (i.e., the Global North). In 2011, for example, about 16.5% of Global North workers worked part time compared to 12% in 2000 (OECD 2012: 239). As with temporary work, though, the extent of part-time work varies significantly. In the European Union, for instance, a fifth of workers are employed on a part-time basis, although the proportion ranges greatly within it, from nearly 50% of all Dutch workers to fewer than 6% in each of Bulgaria, the Czech Republic, Croatia, Hungary, and the Slovak Republic (Eurostat 2015). Likewise, about 20% of workers in the United States work part time (US Bureau of Labor Statistics 2015a). In Japan, overall about 20% of workers are part-timers, while in Australia it is 25%, in Colombia 17%, in France 14%, in Germany 22%, in Ireland 23%, in South Korea 11%, in Poland 7%, in Russia 4%, and in the United

Kingdom 24% (OECD 2016a: 48). In most Global North countries the proportion of women who work part time is higher than is the case for men (though there are wide disparities). Hence, while only about 20% of Dutch men are part-timers, over 60% of women are, largely because the Dutch social welfare system makes it relatively easy for mothers to work part time as they raise their children. However, whereas many part-timers take such jobs because they fit with their desire for flexible lifestyles, the number of workers in the OECD countries (essentially, the world's thirty-five richest countries) who have been taking them because they have no other choices has been increasing in recent years, almost doubling since 2000. This highlights the growing precarity being felt by many workers in these countries. Although, then, women are typically more frequently employed in part-time jobs than are men, in some countries it is men who are increasingly taking such work, whether they want to or not. Generally, women living in countries where they have more opportunities (for example, in many EU nations, where there exist good social support systems that enable them to balance motherhood with working outside the home) are less likely to work part time without having chosen to do so; for men, on the other hand, the predominant factor leading growing numbers to work part time involuntarily is the loss of access to the full-time jobs in which they have traditionally been employed (OECD 2016b).

A third type of work that has grown in recent years in Global North economies is that of self-employment/ freelancing. Although in some cases this is undertaken by people who have quite secure economic futures – for example, accountants who open their own business – for millions of others this type of arrangement has been forced upon them and can leave them in quite precarious employment situations. Although exact numbers are difficult to come by, in 2011 there were about 32.8 million people (15% of the labor force) in the European Union who were considered "independent workers," a number that has been increasing in recent years. Drilling down a little on this number, the European Forum of Independent Professionals estimates that, in 2011, within the EU there were 8,569,000 "I-Pros" ("Self-employed workers, without employees, who are engaged in an activity which does not belong to the farming, craft or retail sectors [and who] engage in activities of an intellectual nature and/or which come under service sectors" – a definition that largely excludes millions of "independent workers" toiling in manufacturing, retailing, and agriculture), with Italy, the UK, France, and Germany home to 5.5 million of these (Rapelli 2012). The bulk of these workers were in scientific and professional fields, social work, and the information and communication sectors. Moreover, I-Pros are now the fastest growing segment of the EU labor force. For its part, in the United States the rise of "on-demand" companies like Uber – founded in California and now operating in more than sixty-five countries worldwide – and the restructuring of work by firms that want to divest themselves of responsibilities toward their employees by turning them into "independent contractors" has helped the number of such workers grow significantly. Whereas in 2015 there were about 3.2 million US "on-demand" workers, this is projected to grow to 7.6 million by 2020 (Intuit 2015). Overall, there are presently about 53 million "freelancers" in the United States (34% of the labor force), including some 21.1 million "independent contractors," 14.3 million "moonlighters" (people who do work additional to their main job – perhaps working for a design company but doing private design work on the side), and 9.3 million "diversified workers" (people with multiple sources of income – for instance, working as a dental assistant and also driving for Uber on the weekend) (Freelancers Union n.d.). In South Korea, for example, a significant development has been the growing use of in-house "independent contractors." Although industries like construction, shipbuilding, automobile assembly, and steelmaking have used in-house subcontractors for several decades, after the 1997 Asian Financial Crisis the practice became much more widespread, such that by 2013 about 46% of Korea's 18.2 million workers were contingent (Jeon 2014a; 2014b).

Although independent contracting has been growing in North American and Europe, this is not, however, a universal phenomenon. Rather, its prevalence reflects the underlying dynamics of different societies. In Japan, for example, the historical dominance of the so-called "job for life" model of employment means that independent contracting is still relatively uncommon, although growing numbers of Japanese have involuntarily adopted it as the economy has suffered economic turmoil since the early 2000s. Likewise, it is also relatively uncommon in China (at least officially), as government regulations tightly circumscribe its use. For its part, in Australia fewer than 10% of the labor force are independent contractors, and the bulk of them are men and older workers (ABS 2015).

In sum, precarious work is increasing across the Global North and has always been common across the Global South (though even those Global South workers who were fortunate enough to secure "standard work" have also felt precarity's grip in recent years). Whether because many workers have had to involuntarily accept part-time or temporary work, because many firms have converted "employees" into "independent contractors" (even as they continue to work in the same places for the same bosses), or because of the growth of many forms of work that appeared to have been consigned to the history books (sweatshops, forced/ indentured labor, industrial homework, and day labor), millions of workers worldwide have seen their lives become more precarious as their connection to the labor market has become looser in recent decades. Without significant opposition to the forces that are driving precarity, this situation is unlikely to get any better. Indeed, even in places where rapid industrialization in recent years has held out the prospect of a regular income for millions of peasants who saw moving from the desperate straits of rural living to work full time in factories as a step up in life, the situation is bleak, as many of these industrial jobs look set to disappear. For example, in China, which overtook the United States in 2010 to become the world's largest producer of manufactured goods, the city of Dongguan's provincial government is spending some US$150 billion to replace workers with robots as a way to address labor shortages brought on by broader demographic changes (see Chapter 2) and also to counter the rising cost of Chinese labor. Such robots, though, will also allow manufacturers to reduce product defect rates and dramatically increase production capacity. Indeed, China is now the world's largest market for industrial robots and Dongguan's "Robot Replace Human" program aims to have completely automated hundreds of factories by 2020.

The rise of precarity and the sense of frustration and hopelessness which often goes along with it – especially on the part of those who have traditionally enjoyed greater security but now fear losing it (such as older white men in the United States and parts of Europe) – is having significant economic, social, and political consequences. In some places, it is encouraging support for nationalist politics and the blaming of immigrant workers for countries' economic problems; in this regard, one can think, among other examples, of the 2016 US presidential campaign of Donald Trump, of those in the United Kingdom who voted in the June 2016 referendum to leave the European Union, and of Marine Le Pen's neo-fascist Front National in France. The rise of precarity also promises to divide societies along other lines, like those of age. Thus, within Europe, the dramatic growth of precarious work is leading to a situation in which many older workers struggle to hold on to the more secure work that they have enjoyed during their working lives, while many younger workers just coming into the labor market know that they will likely never have the economic security enjoyed by their parents. This foreshadows significant issues for the future. In particular, as employers make ever greater use of younger workers on a contingent basis, they have little incentive to train them because they know that they will not have a long-term employment relationship with them. This can lead to skill shortages within the labor force, with consequences for workers (who may not be able to acquire sufficient skills to move into more permanent employment, should it become available), for firms (which, paradoxically, often complain about not being able to find workers with sufficient skills to take the jobs they have on offer), and for societies collectively.[3] At the same time, governments in Spain, Portugal, Italy, France, Greece, and elsewhere, which are concerned about unemployment and about the unwillingness of firms to hire workers (especially young people) because it is too difficult to fire them at a later date, have tried to implement labor law "reforms" that make it easier and less costly to unload workers. The goal, they hope, is to encourage firms to start hiring now if they know that they can easily fire these workers later on should they become surplus to their needs. However,

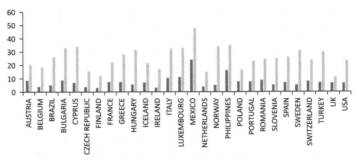

Note: For European countries, the distinction is between permanent and temporary employees; for all others, it is between formal and informal employees. "Latest year available" ranges from 2009 to 2013. depending upon country.

Figure 4.2.   Percentage of households living in poverty by type of employment, latest year available

Source: ILO 2015: 48

although through their actions these governments hope to make employers more willing to hire such workers, these policies ironically simply bolster precarity.

Precarity is also reinforcing social divisions among workers. Thus, permanent employees earn considerably more than do their non-permanent counterparts in most countries, with low wages driving many of the latter into poverty and/or keeping them there (see Figure 4.2). In many countries, this gap between standard and non-standard workers has been expanding in recent years as non-standard workers have seen their living standards fall, although in others it has narrowed as wages for standard employees have slumped due to ongoing economic crisis and restructuring, especially during the 2008 Global Financial Crisis and its aftermath (see Figure 4.3).

Although precarity, then, clearly affects workers in the here and now, it is also important to understand how other ongoing changes being forced upon many working people

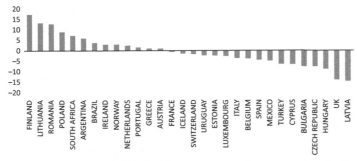

Note: For European countries, the distinction is between permanent and temporary employees; for all others, it is between formal and informal employees. "Latest year available" ranges from 2009 to 2013, depending upon country.

Figure 4.3.   Percentage change in the ratio of average annual wages of temporary/informal employees to those of permanent/ formal employees, mid-2000s to latest year available

Source: ILO 2015: 42

(and not just those who work for low wages) will affect their economic security in the future as well. For instance, in those countries where workers are lucky enough to receive a pension (which, of course, excludes many in the Global South), millions have seen these pensions become less secure as firms and governments have moved away from defined benefit (DB) schemes to pensions with defined contributions (DC). Whereas the former provide pension payments using a set formula (usually based upon income and employment longevity), the latter require workers to invest in financial markets, such that a worker's actual pension payments will rely upon the skill and luck of the individual investor over his/her working life. In Australia, Canada, the United Kingdom, and the United States, this shift from DB to DC pensions has been ongoing over the past three decades or so – in the United States, for instance, between 1980 and 2008 the proportion of private-sector workers participating in DB plans dropped from 38% to

20%, whereas the proportion participating in only DC plans increased from 8% to 31%. Although DC plans do allow some workers to do well, research in the United States has shown that, on balance, there will be more losers than winners and average family incomes of retirees will decline as more workers are placed in DC plans (Butrica et al. 2009). This loss of retirement income is encouraging some older workers to remain in work for longer, meaning that jobs for younger people just entering the labor market are not opening up to the degree that they may otherwise have done. This forces many to take temporary or part-time work because the jobs occupied by people in their parents' or grandparents' generations are not being vacated.

Summary

Significant changes are occurring in the nature of much employment across the globe, changes driven by globalization and labor market neoliberalization. In the Global North, forms of employment that were assumed by many to have gone the way of the dodo – sweatshops, piecework, super-exploitative working conditions – have been expanding significantly in recent years, leaving millions of workers with less secure futures. Arguably, the logical end-point of such developments has been the rise of zero-hour contracts. These contracts are perhaps the perfect expression of employer flexibility, for they allow employers to require workers to be available at a moment's notice but without offering them any guarantee of work – a worker might wait around for twelve hours during any given day but only get paid for fifteen minutes of actual employment. The popularity of such contracts with employers – in 2014, they covered over 10% of the UK's working population – is growing because they offer the ultimate in a flexible labor force. For some right-wing economists and politicians, zero-hours contracts are seen as being highly liberating, fueling individual responsibility on the part of workers by shifting their perceptions of the economy from one in which employers are thought of as the economic parties who are supposed to be "job creators" (and so who have some responsibility to craft a social contract with the population needing jobs) toward a model in which individual workers become "self-sufficient" in the labor market (a mentality which, if internalized, leads them to have no political or economic expectations that firms do things for the benefit of the broader society). Such flexibility, in other words, has a decidedly Dickensian feel to it.

In the Global South, meanwhile, rapid industrialization has brought previously undreamed of wealth for many workers, but under conditions that are reminiscent of the dark satanic mills of late eighteenth-century Britain, when the industrial revolution transformed the country's landscape and led to millions of workers being packed into squalid conditions in urban areas where the new factories were sprouting. However, millions of other working people still toil with little economic security, and even those who have streamed into the factory zones across the Global South are seeing large numbers of jobs now disappear through automation. Combined with significant environmental degradation in many places, the developments described above are threatening to make the future for many people living in such places highly precarious indeed.

CHAPTER FIVE

From Drudge Work to Emancipated Workers?

Beginning in the 1990s a narrative emerged among academics, business leaders, and other prognosticators suggesting that we were crossing a political-economic horizon from an "Old Economy" of dirty, labor-intensive economic activities to a "New Economy" of salubrious "knowledge work." This New Economy, commentators averred, would use technology to liberate workers from the drudgery of traditional jobs. Its leading sectors would be services (including personal services) and "creative" work, with many of the firms involved in such activities being largely online presences rather than having significant investments in traditional brick and mortar facilities. Although manufacturing was still seen as important in the New Economy, it was viewed as experiencing a fundamental transformation in nature. Hence, in the Old Economy the standard model of manufacturing was one in which machinery that could do only one type of thing was used to produce large quantities of standardized goods, a model of production that gave customers little influence over how products were designed – in his 1922 book *My Life and Work*, for instance, Henry Ford quipped that his customers could choose to have their car painted in any color, as long as that color were black. In the New Economy, on the other hand, technology would allow firms to manufacture small batches of individualized goods, frequently on demand, with consumers able to customize the products they bought. Consequently, the form of products would no longer be solely controlled by their manufacturers but could be significantly shaped by the consumer – as when people customize a computer or car by choosing different add-ons. Perhaps the ultimate example of this has been the spread of 3D printing, which American economist and futurist Jeremy Rifkin has suggested is heralding the emergence of a third industrial revolution, one that will replace the assembly-line forms of production that have dominated manufacturing for the past century. These new forms of production, facilitated by technology that allows machines to be easily programmed to perform different operations rather than just a single one, are focused not upon achieving economies of scale (wherein efficiencies are secured by volume of production) but economies of scope (wherein efficiencies are secured by variety of production).

This supposed shift from an Old to a New Economy has many implications for labor and how the labor process is structured. For instance, as detailed in Chapter 4, new flexible ways of organizing manufacturing typically require new ways of organizing labor flexibly, both inside the workplace (e.g., implementing team work) and outside it (e.g., outsourcing). Equally, new forms of work organization, it is suggested, need new forms of management. At least one consultant (Heller 2004) has argued that management in the New Economy must focus upon what he calls the "Five Fs" – it needs to be fast, flexible, focused, friendly, and flat, attributes not often seen in Old Economy management styles, which were frequently bureaucratic, hierarchical, inflexible, ponderous, and therefore slow to respond to market changes. These and other associated developments have led Lazonick (2009) to argue that we have witnessed the death of the "Old Economy business model" (OEBM), one marked by a focus on horizontal diversification and vertical integration to ensure that firms can secure and manage the quality and quantity of critical raw materials and intermediate goods that they require, and under which already successful firms have tended to routinize activities and build upon their superior capabilities in existing product markets to move into new ones. In its place has come a "New Economy business model" (NEBM), in which a manufacturing firm's production increasingly relies upon managing a highly integrated global production network (see Chapter 3). In this scenario, assorted parts of a product are fabricated in different locations and brought together in one place for final assembly. Here, mental and manual labor are reintegrated in the production process so as to take advantage of workers' intelligence and problem-solving capabilities (labor as subject), and, instead of believing that there is a single "best way" of doing things and/or that a product can reach an ultimate level of design, the firm is engaged instead in continuous innovation of both what it produces and how it produces it (the ever more frequent upgrades to computers and iPhones being prime examples). In financial terms, NEBM companies tend to be much more reliant upon venture capitalists for supplying funding to develop new products than are OEBM firms, which have usually drawn upon saved funds or the traditional banking system. Finally, NEBM firms stress interfirm labor mobility rather than any long-term commitment to employees – the NEBM, in other words, marks the end of the "organization man" who stereotypically went to college and then got a long-term, well-paying job with a large firm like IBM or General Motors and retired at the end of a forty-year career with a substantial defined-benefit pension and medical coverage.

Like most narratives of economic change, however, the above is partial and selective. Whereas it is used to suggest that the move from the Old Economy of dreary and dirty drudge work in the extractive and manufacturing sectors to the new, postindustrial economy of wholesome service sector and knowledge work conducted by self-actualizing employees represents a clean break with the past, in reality the past was never quite so focused upon extractive industries and manufacturing as is suggested, and the present is not quite so focused upon healthy postindustrial service and emancipating knowledge work in the way in which it is frequently represented. For instance, looking historically, the 1841 British Census recorded that manufacturing employed 36% of the labor force, while the service sector (including those in domestic service) employed 33%, with half of all employed women and 5% of all men working as domestic servants. In fact, such was the level of service work in early twentieth-century Britain that the 1911 Census counted 1.3 million domestic servants but only 1.2 million agricultural workers and 971,000 coal miners. Of the total working population of 18.4 million, only a little over 6 million labored in manufacturing. For its part, in the United States in 1939 there were as many domestic servants as there were employees of the railroads, coal mines, and automobile industry combined (Stigler 1946: 2). Likewise, manufacturing and the extraction industries are still central parts of the present economy, as witnessed by the miles upon miles of factories that stretch along the US–Mexican border – the (in)famous *maquiladoras* – or across China's Pearl River Delta in Guangzhou province, or the mining operations that suck in hundreds of thousands of workers across southern Africa or in Western Australia. Despite the claim that we have now collectively transitioned into a New Economy in which people are liberated from the alienating grind of mundane menial manual labor, work for millions of people in these and other places is still frequently dirty, highly exploitative, and evocative of nineteenth-century conditions – workers in China's factory zones, for instance, are habitually over-worked to the point that some resort to suicide as a way out of their situation. Meanwhile, whereas in 2013 US manu-facturing directly employed only about 9% of all workers (some 12 million), it had a gross output of US$5.9 trillion – 35.4% of Gross Domestic Product – and supported an additional 17.1 million workers indirectly, more than a fifth of total employment (Scott 2015) – a situation hardly charac-teristic of a postindustrial society. In fact, manufacturing is still by far the most important sector of the US economy in terms of total output and employment, and much of it still uses Old Economy-style Taylorist assembly lines.

In this chapter and in Chapter 6, then, I present brief case studies of work in some of those fields of employment that are imagined to be Old Economy sectors and in some that are imagined to be New Economy sectors. In so doing, I want to detail some of the employment experiences of hundreds of millions of working people at the beginning of the twenty-first century. Clearly, there are other eco-nomic sectors and types of work that I could have chosen to explore and the ones presented here are not meant to be exhaustive. Instead, I present them because they provide particular insights into the nature of work and labor's con-temporary experience across a variety of occupations and places. In so doing, I stress the continuities between work in the supposed Old and New Economy sectors, rather than focus upon their imagined discontinuities.

Laboring in the Old Economy

On the swing to the cancer in the bush: iron ore mining in Western Australia
Iron is the world's most commonly used metal and is central to numerous industries. Australia is home to the

world's largest iron ore reserves – some 54 billion tonnes (29% of global totals). Over 90% of these reserves are located in the Pilbara region of Western Australia (WA) (see Figure 5.1). Production in, and export from, the region first began in earnest in the 1960s. As with most iron ore mines, the Pilbara's are open cut. Using big diggers and

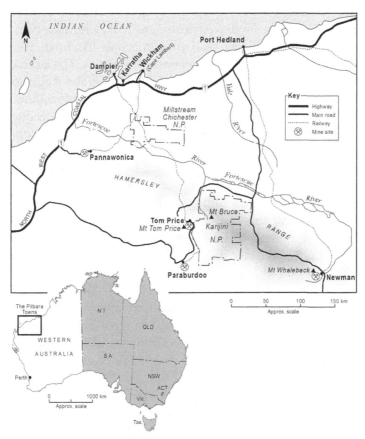

Figure 5.1   The main mining sites of the Pilbara

Map by Peter Johnson

other massive machinery, workers carve ore out of the earth and load it into equally huge trucks (some of which can carry 300 tonnes) for transport to crushing and screening plants. After initial processing, the ore is transported by train, some made up of more than 250 cars and more than two kilometers in length, to ports for export. Each train can carry loads of more than 25,000 tonnes.

The Pilbara itself is frequently imagined to be a remote place. Certainly, its dry environment makes human habitation difficult – fewer than 50,000 people live in an area greater than 500,000 square kilometers in size. The largest town – Port Hedland – has a population of only about 15,000 and is more than 1,600 kilometers (an eighteen-hour drive) from Perth, the state capital. Nevertheless, despite appearances, the Pilbara is highly integrated into the contemporary global economy. It has seven of the world's ten largest mines and their output helps make Australia the world's second-largest iron ore producer – in 2015, Australia mined 824 billion tonnes of ore (25% of world totals), behind only China (1.4 billion tonnes: 42% of world totals), and ahead of Brazil (428 billion: 13%), India (129 billion: 4%), and Russia (112 billion: 3%). The bulk of the Pilbara's output is exported to China. In fact, of the 968 million tonnes of iron ore imported by China in 2015 (making it the world's largest importer), approximately 607 million (64%) came from Australian mines. These imports helped fuel China's massive steel industry, which produced 803 million tonnes of metal – about half of global totals – in 2015 (the next largest producer, the European Union, produced 10% of global totals). Not only is the Pilbara thoroughly tied into the global economy in terms of exports of iron ore, much of which eventually ends up in North America or Europe in Chinese-manufactured goods, but it is also deeply connected via lines of authority and capital flows – of the Pilbara's two largest mining firms, one (Rio Tinto) is headquartered in London, while the other (BHP Billiton) has long roots in London and The Hague. More importantly for my purposes here, though, the Pilbara is linked to the broader world through significant long-distance labor flows.

In the early days, the Pilbara's mining regions were union spaces where thousands of miners toiled. They were also very masculine spaces – the mineworkers were virtually all male. Typically, miners lived in communities close to where they labored, in towns established by the mining companies. The region's labor force was essentially created from scratch as the industry was developed, for prior to mining there were very few people living there except for members of local Aboriginal groups who, because they were not white, were largely overlooked as a labor source anyway (Ellem 2015). Many workers came from other communities in Australia that had long histories of mining and thus of unionism, attitudes which they brought with them to the Pilbara. The type of labor relations that developed in the area were shaped by broader state- and national-level labor laws which set minimum standards for wages and working conditions under the "award system" of compulsory arbitration that had developed in the early twentieth century and which, at its height, covered about 85% of the Australian workforce (Plowman et al. 1980). One key feature of work in the Pilbara, though, is the geographical dispersal of the mines. Whereas in coal mining regions like the north of England or Appalachia in the United States it is common to have several mines close together, in the Pilbara each company's mines are often hundreds of kilometers apart. This posed logistical problems for the companies but presented opportunities for the union – although the distances between mines made it difficult for workers to organize across space, the communities' largely self-contained nature meant that within them the union could easily dominate most aspects of town life. It also meant that it had significant control over the work process and was thus able to secure wages well above the national average. Power in the workplace and power in the community, then, were mutually reinforcing.

While the mining communities had first developed largely as union spaces, in the 1980s the companies began to develop strategies to challenge worker power. Two developments were crucial. First, the regulatory environment within which labor relations were conducted changed, largely under pressure from the kinds of neoliberal ideas discussed in Chapter 4. Among other things, this resulted in the undermining of the traditional award system in which the state – at both the national and the WA level – set a minimum wage and health and safety requirements below which collectively bargained agreements between workers and their employers could not fall. Increasingly, this system was replaced by a model in which the elements of union contracts were largely determined by direct local bargaining between employer and worker representatives. By the 2000s, under the WorkChoices legislation introduced by Prime Minister John Howard's federal government, this had devolved further into a system whereby employers and workers could even negotiate individual agreements. The effective undoing of the award system and the longstanding social safety net that it had provided, together with the introduction of anti-union legislation designed to make labor markets more flexible, significantly undermined organized labor's power in the Pilbara (and across Australia more generally). Second, new ways of organizing the labor force were introduced, principally that of a system of labor supply called "Fly-in, Fly-out" (FIFO) – sometimes referred to as being "on the swing" – in which, instead of living in the towns of the Pilbara with their families, miners live in cities like Perth, Sydney, Melbourne, Adelaide, or even Auckland in New Zealand and Bali in Indonesia, commute into the Pilbara to work for several days at a time, and then fly back to their homes. In addition to the actual miners, there are also tens of thousands of construction workers and other FIFO workers engaged in surveying, driving goods back and forth, and working at plant maintenance and work camp operations. This system has also sometimes been supplemented by DIDO (Drive-in, Drive-out) labor sourcing.

FIFO is certainly not a new form of supplying labor to remote, resource-extracting areas. Its first large-scale use occurred in the 1950s in the Gulf of Mexico's offshore oil sector, where establishing permanent communities was clearly not possible. By the 1980s, it had been introduced into WA's mining industry and in 2015 some 60,000 individuals were FIFO workers, toiling in iron ore, gold, and other extractive industries (Parliament of Western Australia 2015: i). Whereas in 1990 approximately 19% of the WA mining industry's workers were employed as FIFOs, by 2000 that had increased to 31% and to 49% by 2005. At that time, about AU$12.6 billion per year of mining operations in WA were reliant upon FIFO operations, which was 47% of the income that came from mineral and petroleum operations in the state (Western Australian Local Government Association 2010: 23). In the Pilbara itself, there were 13,250 resident mineworkers and 6,700 FIFOs in 2008; by 2010, the numbers of residents and FIFOs were almost even: 15,900 and 15,400 respectively (Western Australian Local Government Association 2010: 17–19). Five years later, 57% of Pilbara mining employees were working as FIFOs (Morris 2012: 7). In 2011, of the 22% of Australia's mining industry workforce that usually resided in Perth (many of whom were executives and others who never went anywhere near an actual mine), about 16% worked in the Pilbara and another 7% worked in WA's Goldfields-Esperance region (ABS 2013). Assuming that each FIFO worker has three direct family members, this means that some 240,000 Western Australians (nearly 10% of the population) were directly impacted by FIFO.

The mining companies have suggested several reasons for shifting to FIFO work. These include avoiding the cost of building and operating new mining towns (the Argyle diamond mine in northeast WA, for instance, estimates that it saved AU$50–70 million in construction costs by using FIFO; see Storey 2001), avoiding long lead times and environmental consequences for approving and building new towns, and allowing miners to live in cosmopolitan places like Perth and Sydney rather than in isolated mining communities in the middle of nowhere. Others have been more open about how it saves them on labor costs – the Fortescue Metals Group has estimated that it would cost them an additional AU$100,000 per worker per year (largely in housing costs) to employ miners *in situ* rather than as FIFO workers (*Sydney Morning Herald* 2012). Perhaps one of the most significant reasons, however, has been the way in which the mining companies have been able to use FIFO to break the industry's unions and turn the Pilbara into a space which is much more acquiescent in terms of company goals than it had been in the past. In particular, by switching to FIFO operations, companies have been able to fragment social networks between workers that were the basis for powerful unionism. Hence, when workers lived and worked closely together, they developed profound interpersonal relationships outside the workplace which served as a bedrock for union activities within it. However, with the shift to FIFO many miners now live hundreds of kilometers away from the mines and hundreds if not thousands of kilometers from one another. The undermining of their social networks outside the workplace as a result of FIFO, then, has weakened worker solidarities and made it easier for the mining companies to force concessions on their labor forces. Moreover, the intensity of FIFO work is such that, even when they are in the mining areas, miners' shifts are so exhausting that they seldom have much time to talk about the union; grueling 12-hour shifts, when summer temperatures frequently hit 45°C/113°F, leave workers simply too tired to do much but collapse into their bunks at the end of the day before it all starts again the following day. By transforming the geographical relationships between miners outside the workplace and between miners and their places of work, the mining companies have used spatial divisions to weaken the unions and reassert control over the Pilbara's production spaces.

More recent innovations threaten to take such control one step further and perhaps even to automate the industry to the point where miners are no longer needed – whereas China's "Robot Replace Human" plan, discussed in Chapter 4, is leading to workerless factories, then, automation is potentially leading to a workerless Pilbara. In particular, companies have been experimenting with employing "driverless" trucks and trains, machines operated via remote-control using video-game-type technology similar to that utilized by the US air force when conducting drone strikes in Afghanistan.[1] For instance, as part of its "Mine of the Future" program in which "People and technology [will be] working together" (Rio Tinto 2014) – a very New Economy-sounding proposition – Rio Tinto is spending AU$518 million to implement its AutoHaul project, the world's first fully autonomous heavy-haul, long-distance

rail system (*Financial Times* 2016). For its part, the
Fortescue Metals Group has installed autonomous trucks
at its Solomon mine (*Australian Mining* 2013). Controlled
remotely by operators sitting in air-conditioned offices in
Perth, such equipment is portrayed by the companies as
being more efficient, safer and cleaner for workers, and
cheaper to operate than is using human miners, largely
because the machinery has less downtime. It is entirely
likely, however, that the jobs of operating these machines
will someday be outsourced from Perth to cheaper labor in,
perhaps, India or pretty much anywhere else in the world,
thereby undermining the long-held belief that the job of
digging resources out of the ground is geographically constrained to the places where those minerals are found.

In surveying these developments, it is also important
to consider the stresses that FIFO practices can place on
individual workers and their families. Miners being away
from home for considerable periods of time often fractures
households and the "4/1 swing" (four weeks of work followed by one week off) is commonly called the "divorce
roster" in the industry. Likewise, research suggests that
separation anxiety experienced by FIFO workers' children is frequently expressed in their engaging in socially
destructive behavior, while loneliness experienced by FIFO
and DIDO workers can manifest itself in substance abuse
and mental illness. Indeed, the stress of FIFO work and
living in isolated work camps, separated from family and
constantly working, has led some workers to suicide. A
recent inquiry found that between 2008 and 2013, twenty-
four WA workers committed suicide. Seven of these were
explicitly named as FIFO workers, while the others worked
in occupations that suggested they were also FIFO workers
(*Sydney Morning Herald* 2015). Rather than liberating workers, as the New Economy evangelists argued would happen,

the new labor control systems employed by Australian mining companies appear to be having severely deleterious effects upon them, to the extent that the mayor of the gold-mining town of Kalgoorlie has called the FIFO/DIDO practice the "cancer of the bush."

Sweet work? Cocoa plantation workers in West Africa
The average American consumes 5 kilograms of chocolate annually. Compared to many other nations, this amount is paltry: the average Swiss consumes 9 kilograms, the average German 7.9, and the average Briton about 7.4 (McCarthy 2015). In fact, the ten countries with the largest per capita consumption are all European. In 2010, the chocolate industry was worth about US$83 billion a year, greater than the GDP of more than 130 of the world's countries. Europeans ate almost half of that chocolate; Americans consumed 24%. Asia, South America, and Africa accounted for only 15%, 9%, and 3% respectively (Afoakwa 2016: 35). However, chocolate sales in Asia have been increasing recently and China is expected, before too long, to be the world's second-largest consumer, after the United States. Although the cocoa bean that is used to make chocolate is indigenous to Latin America, more than 70% of world cocoa production occurs in Africa, to which the cocoa plant was introduced in the nineteenth century (see Table 5.1). Cocoa is predominantly a smallholder crop, and over 90% of production occurs on small farms; in West Africa and Asia, the typical farm is only about two to five hectares in size. In Africa, one hectare produces about 300–400 kilograms of cocoa beans annually, while in Asia the figure is usually about 500 kilograms. Total global production increased by 13% between 2008 and 2012, from 4.3 million tonnes to 4.8 million (World Cocoa Foundation 2014: 2–3). Although cocoa is a tropical plant, the bulk of

Table 5.1   Production of cocoa beans 1970–2013 by country, ten largest producers (thousands of tonnes)

Country	1970	1980	1990	2000	2010	2013
TOTAL AFRICA	1,121	1,026	1,522	2,351	2,784	3,015
Cameroon	134	117	115	123	264	275
Côte d'Ivoire	179	417	808	1,401	1,301	1,449
Ghana	406	277	293	437	632	835
Nigeria	305	153	244	338	399	367
TOTAL AMERICAS	381	553	555	462	630	721
Brazil	197	319	256	197	235	256
Dominican Republic	38	28	43	37	58	68
Ecuador	54	91	97	100	132	128
Mexico	29	36	44	28	61	82
Peru	2	4	15	25	47	71
TOTAL ASIA	116	585	410	509	880	801
Indonesia	2	10	142	421	845	778
TOTAL WORLD	1,543	1,671	2,532	3,373	4,341	4,586

Source: FAOStat 2016

cocoa processing takes place in the Global North (though Indonesia and Malaysia are also important, processing both local beans and those from West Africa) (see Table 5.2).

As they anticipate its melting in their mouth, few people probably think much about the labor processes involved in bringing a piece of chocolate into their hand. However, the chocolate that a North American or European consumer may relish is embroiled in wide-ranging GPNs that link Global South producers of the humble cocoa bean with, mostly, Global North TNCs and, then, consumers across the planet. Such GPNs intersect with several other industries, including: milk production; sugar, fruit, and nut growing (chocolate manufacturers consume about 40% of the almonds and 20% of the peanuts cultivated across

Table 5.2 Grinding of cocoa beans by country, 2009–2012 (thousands of tonnes)			
Country	2009/10	2010/11	2011/12
TOTAL EUROPE	1,530.3	1,624.5	1,521.3
Belgium	70.0	75.0	70.0
France	145.9	150.0	128.0
Germany	361.1	438.5	407.0
Italy	63.2	66.5	66.6
Netherlands	525.0	540.0	500.0
Russia	51.9	60.9	63.0
Spain	86.0	86.0	90.5
UK	110.0	87.0	78.0
TOTAL AFRICA	684.5	657.5	716.5
Côte d'Ivoire	411.4	360.9	430.7
Ghana	212.2	229.5	211.7
TOTAL AMERICAS	814.2	861.5	845.5
Brazil	226.1	239.1	242.5
Canada	59.2	62.3	60.0
US	381.9	401.3	386.9
TOTAL ASIA AND OCEANIA	707.7	794.6	873.5
Indonesia	130.0	190.0	270.0
Malaysia	298.1	305.2	296.8
Singapore	83.0	83.0	83.0
Turkey	68.0	70.0	75.0
TOTAL WORLD	3,736.8	3,938.1	3,956.7

Note: Only countries grinding more than 50,000 tonnes in 2011/12 are shown. Data run from October to September. Figures in italics are from official or trade sources; other figures are International Cocoa Organization estimates based upon trade in cocoa.

Source: International Cocoa Organization 2014

the planet); and metal mining, lumbering, paper making, and dye manufacture, all of which go into packaging. These chocolate GPNs have their roots in a system that is dependent for its profitability upon the labor of millions of children across West Africa and elsewhere. Although children have always been involved in cultivating cocoa, a collapse in prices in the early 1990s and again in recent years (this latter associated with bumper harvests and economic problems in Europe) greatly increased pressures to use children as a cheap labor source. Rising global demand for chocolate as Asian consumers have become wealthier has reinforced these pressures, such that the numbers of children working to produce cocoa increased by 51% between 2009 and 2014. As one producer has explained, illustrating how he sees his labor force's members as essentially disposable: "Children are easier to discipline than adults. They wait and wait and wait and then, at a certain stage, they get fed up and escape and we get new ones" (*Telegraph* 2001).

The ILO (2013a) estimates that in 2012 there were 168 million child laborers in the world (11% of the world's children), of whom 59% worked in agriculture. Although this figure is significant, it is much less than the 246 million who were working in 2000. Of the 168 million, 85 million were employed in hazardous work (compared to 171 million in 2000). While the largest numbers of child workers are found in Asia and the Pacific (almost 78 million, or about 9% of the region's children), sub-Saharan Africa has the highest incidence of child labor – one in five children (a total of some 59 million). Globally, most children work in agriculture (some 98 million), but about 54 million work in services and 12 million in industry, mostly in the informal economy. According to some estimates, there are as many as 2.1 million working in the cocoa fields of Ghana and Côte d'Ivoire. The vast majority of child cocoa workers toil under fairly hazardous conditions, including using sharp machetes (in Côte d'Ivoire 37% of them had suffered wounds), working long hours, carrying heavy loads (children frequently drag 50 kilogram bags of beans through the forest), and/or being exposed to agro-chemicals (Tulane University 2015). Most of these children are between about 11 and 16 years of age, though investigative journalists have found some as young as 5. In many cases, they are so young and malnourished (and therefore small) that the machetes they wield to cut the cocoa pods from the trees are as tall as they are. About 40% of these children are girls. Both boys and girls also sometimes suffer sexual abuse at the hands of their employers.

There are two aspects of cocoa harvesting that I particularly want to highlight with regard to connecting working practices in West Africa with how chocolate GPNs function to bring cheap chocolate to global consumers. The first involves how the labor force upon which the GPN relies is created. In the case of the actual manufacture of chocolate in Europe or North America, the workers involved are typically hired in the same manner as any other factory worker and they do fairly typical factory work. However, when it comes to securing the essential raw material for chocolate, the industry has become reliant not just on children but, increasingly, on the forced labor of children. Hence, although many children are the offspring of the cocoa farmers themselves, many others are trafficked from countries in the West African interior to work, including in cocoa harvesting and preliminary processing (mainly extracting the beans from pod husks and drying them). These are not the only trafficked children in the region. There are boys from Guinea who work in Côte d'Ivoire's mining regions, in construction in Togo, and fishing in Ghana. Many young girls and women are trafficked to work as domestic servants or street vendors. Some are also forced into sex work, both in West Africa and in Europe. Whereas the children trafficked to work in Ghana and Côte d'Ivoire primarily come from other West African countries, those who work in Congo's cocoa fields come from nearby countries like the Central African Republic, Cameroon, and Benin (which provides the largest number of trafficked children to Congo), as well as farther afield (US State Department 2016). It is this trafficking to expand the industry's labor force that has allowed production to keep up with the increased demand for cocoa. Some of these children are kidnapped and forced to work, while others are sold by their desperate parents to brokers or directly to farmers – in some instances, the parents receive cash payments (perhaps US$30–50, the amount the average person in Britain spends on chocolate in three to five months), while in others they receive livestock (for example, a calf for every year that their child works).

The second aspect concerns how cocoa producers immobilize their labor force once it has been assembled. In order to inure new child recruits to the hardships of collecting cocoa, they are often routinely beaten during the first six months (ominously often referred to as the "breaking-in period"), especially if they try to escape their situation – as one child rescued by Ivoirian authorities has reported: "The beatings were a part of my life ... I had seen others who tried to escape. When they tried, they were severely beaten" (*Daily Beast* 2015). Trapped on isolated farms, the children are often worked from dawn to dusk in oppressive heat and fed a diet only of bananas, which leads them to develop vitamin deficiencies. Few manage to escape – what little they earn is often stolen by their employer, so they are unable to purchase a bus ticket home or to pay for the food and other needed expenses (including bribing border guards). Given that they have spent their lives on cocoa farms, virtually none has any education, so opportunities to do other types of better-paying work are few and far between, which makes staying seem like their only option. If they have been trafficked from far away, they generally do not speak either the local African language or the appropriate European language (French or English) and have usually had their identity papers confiscated "for safe keeping" by their employers, which complicates their efforts to move about. Some farmers use belief in magic to control the children, telling them that they have had spells cast on them that will cause them to become paralyzed or worse should they seek to escape. The abuse they suffer frequently causes not only physical injuries but mental scars. Many are psychologically broken and their physical and emotional isolation leads them to become attached to their employers, whom they see as their only protectors. Given this, many have trouble readjusting to being with their families if they do escape. Through these means and others, then, the cocoa farmers create a quite literally captive labor force.

Despite the difficulties involved in trying to improve conditions for these child workers in an industry of countless smallholdings (in Côte d'Ivoire alone there are about 600,000 cocoa farms), some progress has been made. In 2001, several US Senators attempted to require chocolate manufacturers to certify that their products were not made using slave labor. However, powerful industry forces managed to kill the effort. Instead, industry representatives argued for voluntary self-compliance and established some pilot programs to create awareness among farmers in West Africa concerning non-exploitative labor practices. They also introduced programs to provide farmers with improved access to credit so that they can make a living without having to use child labor and sought to establish a system to trace the geographical movement of the cocoa beans as a way to certify that they have not been produced by children. The US Chocolate Manufacturers Association also signed on to the Harkin–Engel Protocol, a voluntary public–private agreement to eliminate child labor's worst excesses. However, the failure actually to implement this protocol put pressure on the companies to do something, and in 2010 they agreed to The Declaration of Joint Action to Support Implementation of the Harkin–Engel Protocol, in which they pledged to reduce the worst forms of child labor by 70% by 2020. Some real progress was made in 2009 when Cadbury Dairy Milk in the United Kingdom became the first major brand to make all its chocolate fair trade certified. The company subsequently extended that commitment to its dairy milk bars sold in Australia, Canada, Ireland, and New Zealand. For its part, in 2010 the US chocolate firm Hershey issued a corporate social responsibility report, though this has been widely criticized as simply a form of "greenwashing" designed to make the public believe that its policies and products are socially and environmentally responsible. In fact, the company – along with Nestlé and Mars – was sued in 2015 by a number of California plaintiffs on the grounds of false advertising for failing to disclose the use of child slavery on packaging. Although the companies all said that the suit was without merit, Nestlé claims to be tackling issues of child labor through a more than US$100 million action plan to establish a monitoring and remediation system. Meanwhile, Mars points to its "Vision for Change" program launched in 2012 that is meant to encourage sustainable production by building 16 cocoa development centers and 52 cocoa village centers in the Ivory Coast, in which farmers are taught improved production and management techniques, although the company does not specify how this will reduce child labor. In the meantime, millions of children continue working on cocoa farms under brutal conditions to secure the raw material that satiates the planet's sweet tooth.

Fishy business: forced labor in the seafood industry

The seafood industry is another in which questions of forced labor have come to the fore in recent years. One reason for this is the fact that, since the 1970s, the deep-sea shipping industry's labor market has become global, especially with the growth of so-called "flags of convenience" (FOCs), whereby owners register their ships in countries like Panama, Liberia, and the Marshall Islands to avoid their own countries' more stringent labor regulations (Yannopoulos 1988). Whereas in 1950 about 4% of the world's fleet used FOCs, today over 60% does so. Even landlocked Mongolia has a thriving FOC registry. Many countries' fleets have more tonnage registered overseas than domestically (see Figure 5.2). The growth in the use of FOCs has allowed owners to dramatically reduce labor costs by shrinking crew sizes, by hiring crews out of low-wage Global South countries, and by turning their employees into "self-employed workers." Not only does the lack of strong labor protections encourage abuse of workers, but

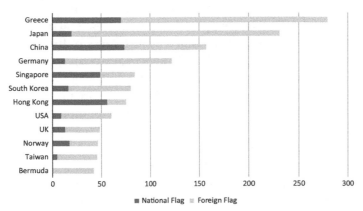

Figure 5.2   Size of merchant fleets, millions of deadweight tonnes, as of January 1, 2015

Source: UNCTAD 2015b: 36

the fact that crews are often made up of workers from four or five different countries means that language barriers can limit their ability to come together to protect their interests. Lack of enforcement of the even minimal labor protections enjoyed by workers results in them frequently having their wages stolen – the extent of the problem is hinted at by the fact that in 2014 alone the International Transport Workers' Federation (ITF) recovered US$59.5 million in wages owed to mariners (Seafarers' International Union 2015). Even countries typically thought of as having relatively strong protections, such as New Zealand, have been rocked by reports of Indonesian, Filipino, Korean, Chinese, and other workers being subjected to labor rights violations – in one week in 2014, ITF representatives recovered US$110,000 in unpaid wages for crews docked in New Zealand ports (Maritime Union of New Zealand 2014). Many of these crews work on boats contracted by New Zealand operators, a phenomenon that has become increasingly common since the late 1970s as Global South workers have provided cheaper labor alternatives to New Zealand workers. As one commentator has observed: "What has happened is that the New Zealand fishing industry is being developed on the backs of foreign ... exploited labour" (quoted in Stringer et al. 2016).

While accounts of seafarers being forced to work on ships under extremely poor conditions – sometimes even at gunpoint – have become increasingly common, so, too, have reports of forced labor on the landward side of the seafood industry. Much of this relates to the globalization of seafood consumption patterns and concomitant changes in the industry. Seafood is one of the most widely traded commodities around the globe, with more than half of all seafood (whether wild caught or farm-raised) originating in the Global South but with nearly three-quarters of all seafood being consumed in Europe, Japan, and the United

States – in the case of the latter, 23% of seafood imported in 2009 came from China, 16% from Thailand, and 16% from Indonesia, Ecuador, and Vietnam (US Government Accountability Office 2011: 6). Some 55% of seafood eaten in the European Union and 85% eaten in the United States is imported. Shrimp is the leading import in both places – about 90% of the shrimp eaten in the United States is imported (Marks 2012). Not only is the United States today importing more of the seafood it consumes than it did a decade ago, but increasingly that seafood is coming from fish farms – about half of all seafood eaten is farm-raised. As imports from Asia and South America have risen, US-based producers have come under significant pressure to reduce costs so as not to lose market share. This has led them to seek to reduce labor expenses, which are a significant component of the cost of processing seafood, much of which must still be done by hand. This has resulted in worsening conditions for many workers. In 2016, the US National Guestworker Alliance (NGA) published a report detailing abusive and exploitative conditions suffered by immigrant workers – some of whom were undocumented and some of whom were working legally as part of the federal government's H2-B guestworker program – who were processing shrimp and crawfish at plants in Massachusetts and along the Louisiana Gulf Coast. Such abuses included employers stealing wages, making their employees work longer than allowed, and providing them with sub-standard housing, all as a way to drive down labor costs to stay competitive with foreign producers. Given their precarious positions, most workers were afraid to complain for fear of losing their jobs and what little wages they earned. Arguably, however, it is the Southeast Asian seafood industry that is now the most emblematic when it comes to issues of forced labor globally.

A number of recent exposés have focused attention on the horrendous conditions experienced by workers in Southeast Asia's seafood industry. Journalists from the *Guardian* (2014, 2016) newspaper in the UK, for instance, have detailed a network of forced labor in both catching shrimp at sea and in farming and processing them on land. This happens throughout Southeast Asia's Greater Mekong Subregion, but it is particularly rampant in Thailand, where it has the potential to worsen in coming years as growing competition in the industry has put even greater pressure on Thai producers to cut labor costs – although Thailand has been the world's largest shrimp exporter since the 1990s, more recently its industry has faced greater competition from Vietnam, Indonesia, China, and Ecuador, among others. Prior to the mid-1980s the vast bulk of Thai shrimp were wild caught, mostly in the Gulf of Thailand. However, growing demand for Tiger shrimp (*Penaeus monodon*) encouraged farming to expand in earnest. Myriad rice farmers converted coastal fields and the mangrove forests that often border them into ponds. Although this has had significant environmental consequences, it has been highly profitable for farmers – with a pond the size of a football field capable of producing about five tonnes of shrimp annually, many farmers who previously might have made US$500 a year can make many times that. Moreover, although shrimp farming of the type practiced for centuries could result in yields of up to 500 kilograms per hectare (*circa* 450 pounds per acre), the modern industrial-style operations can produce nearly 100,000 kilograms per hectare (*circa* 89,000 pounds per acre). Today, about 70% of Thailand's production is farm-raised and the country is the world's third-largest exporter of seafood, with yearly revenues of about US$5 billion. The industry's growth has led to significant increases in the need for labor, much of it from neighboring countries (accounting for more than 90% of those working in Thailand's fishing sector) and much of it trafficked. It has also led to significant environmental damage – in the early 1990s, for instance, hundreds of Thai shrimp farmers illegally moved operations into some of the country's protected wetlands.

For those workers – usually young men or boys – who catch wild seafood, employment frequently involves working on boats that spend months or even years at sea. As a way to reduce fuel expenditures and to ensure that boats can spend as much time fishing as possible, the industry is set up so that smaller boats ferry supplies to the main boats and take their catches back to shore for processing while the fishing crews stay on board. The result is that it is not uncommon for crew members to remain at sea for years. As a way to improve productivity, crew members are often forced to take methamphetamines so that they can work twenty-hour or longer shifts. There have also been reports of workers being murdered and dumped overboard to avoid paying them their wages or if they are too ill to work and if keeping them on board represents a "poor investment" of food and water. Because boats operate far from shore in international waters, monitoring labor abuses is often difficult, a situation exacerbated by poor registration and licensing practices (many boats operate using false papers, for both the boat and the crew). The boats, often using crews that have literally been purchased from various labor brokers, catch many types of seafood, including squid, shrimp, tuna, and so-called "trash fish," which are inedible but which are ground into fishmeal for shrimp farming. One of the biggest consumers of such trash fish is the world's largest shrimp farming company, Thai-based Charoen Pokphand (CP) Foods. Although there is no evidence that CP Foods itself uses slave labor, it and other

companies are nonetheless incriminated in the phenomenon because of the tight connections between the boat
operators and land-based producers and processors in the
shrimp industry's GPN. This, in turn, means that the firms
to which CP Foods sells its products – well-known Western
retailers like Walmart, Carrefour, Costco, Tesco, Morrisons,
the Co-operative, Aldi, and Iceland stores – are also implicated, as are many restaurants and producers of canned
pet food across the globe; investigators have traced Thai
shrimp to the retail stores Dollar General and Petco, as well
as US restaurant chains like Red Lobster and Olive Garden.

One of the forces driving the spread of appalling working
conditions in the industry has been the growth of Western
consumers' desire for ready-to-cook shellfish, which must
be peeled prior to being packaged and exported. The de-
heading, peeling, and de-veining of shrimp is the most
labor-intensive part of their processing and, in a desire
to keep labor costs as low as possible, many exporting
companies have subcontracted this work out to small local
firms operating "peeling sheds." As of 2011, there were
about 200 such sheds in Thailand formally registered
with the government, and many hundreds more unreg-
istered ones (Environmental Justice Foundation 2013:
5). The operations of registered, and especially unregis-
tered, sheds go largely unregulated yet employ hundreds
of thousands of migrants, both Thais themselves but
also people from Burma, Cambodia, and Laos, who have
often been smuggled into the country illegally. In Samut
Sakhon, Thailand's main seafood processing region, anti-
slavery activists estimate that in 2011 there were more than
400,000 migrant workers from Burma alone, about one-
third of whom had been trafficked. Workers may earn as
little as US$4 a day for peeling 80 kilograms (176 pounds)
of shrimp (Associated Press 2015). Once the shrimp have

been processed at these sheds, they usually then head off for secondary processing – such as breading and packaging – before being exported. Whereas peeling sheds are generally run by small-scale firms, the secondary processing is typically conducted by larger-scale operators. Although these latter operations are better regulated than are the peeling sheds, many of the same types of labor abuses can nevertheless be found in them.

Finally, developments in other fields are also driving demand for shrimp. In particular, a growing desire for shrimp byproducts has placed pressure on Thai producers. For instance, shrimp and crab shells contain the polymer chitin, which has several industrial uses, including as a thickening agent in the food industry and for hydrating in the cosmetics industry. In agriculture, chitin is used both as a fertilizer and as a chemical that can induce defense mechanisms in some plants, thereby protecting them against disease, while fruit-growers spray it onto fruit to prevent water loss. Chitin is also used as a binding agent in various dyes, fabrics, and adhesives and can be used to strengthen paper. In addition, it has been used as a form of biodegradable plastic and shows promise as a substrate for engineering human tissues. Moreover, its flexibility and strength mean it can be made into surgical thread, while its biodegradability means that it will dissolve in flesh. As a result of these uses, the global chitin market was estimated to be worth about US$1.35 billion in 2013 and is likely to grow significantly in the future, given that every year about 1.5 million tonnes of waste crab, shrimp, and lobster shells are produced in southeast Asia. Although dried shrimp shells are worth only about US$100–US$120 a tonne, once processed into high-end chitin they can be worth as much as US$200,000 a tonne. This has encouraged the establishment of more than twenty Thai companies producing chitin-related products, mainly for agricultural applications. The need for shells as a raw material, then, will likely place additional pressures on seafood producers in Thailand and neighboring countries not only to expand production, but also, in all likelihood, to maintain or even exacerbate the industry's reliance upon cheap, frequently trafficked, labor.

Certainly, all is not hopeless for the workers involved in Thailand's shrimp industry. In 2013, some 500 Burmese migrant workers went on strike at CP Food's Rayong factory to protest their working conditions, expensive and hidden agents' fees, and the fact that their employer had confiscated their passports (Environmental Justice Foundation 2013: 17). Equally, as with chocolate so with shrimp, legal avenues have been pursued as a way to improve workers' situations. Several California law firms have filed a class action lawsuit against Costco and its Thai seafood supplier for allegedly knowingly selling shrimp produced using slave labor. In 2016 Cambodian villagers filed suit in a US District Court against two Thai companies (Phatthana Seafood and SS Frozen Food), which are major exporters to the United States, and two US importers (Rubicon Resources LLC and Wales & Co. Universe), claiming that the defendants were knowing participants in a joint venture that profited from the import and sale of seafood produced with trafficked labor. Although some of these firms' operations occur outside the United States, the plaintiffs' attorneys sued under a US law that holds liable anyone involved in any part of a supply chain that brings into the country goods produced using slave labor. In 2012, the US government also prohibited federal agencies from purchasing shrimp produced by forced labor, although the fact that slave-produced seafood can easily be mixed with other fish and crustaceans means that keeping it out of the US food supply is more complicated than simply boycotting farms known to be using trafficked workers.

Summary

Despite claims that we have moved from an Old to a New Economy way of organizing work, one in which work is clean and workers are emancipated from drudgery, it is clear that for millions not only has work not become less dirty and exploitative but, in fact, it has become more so. For instance, although mining in the Pilbara does not involve forced labor and miners are well paid, pressures to produce ever greater quantities of iron ore under extreme conditions are pushing many to the brink, to the point where several have killed themselves. Equally, while slavery (often also referred to as "bonded labor," "forced labor," "indentured servitude," "restavec," or "attached labor") is frequently thought of as a labor practice that died out in the nineteenth century thanks to the abolitionist activities of people like William Wilberforce and the defeat of the Confederacy in the US civil war, it is alive and well in the era of the New Economy. Indeed, there are an estimated 20–45 million slaves across the planet, of whom up to 500,000 are believed to work in the United States, in the commercial sex industry, in agriculture, in domestic service, and also in manufacturing. Slavery is as central to certain modern global supply chains as it was to the eighteenth- and nineteenth-century sugar trade – it can be found in the production of cocoa and seafood, the growing of Brazilian soy beans used as feed by Western agribusinesses producing chickens, the free-range production of eggs used by UK retailers like McDonald's, Tesco, Asda, Marks & Spencer, and the Sainsbury's Woodland brand, and the cultivation of all-year salad crops in Spain, to name just a few. The last country to officially outlaw slavery – Mauritania, in West Africa – did so in 2007. Nevertheless, slavery's continued prevalence can be seen in the fact that some countries have recently felt the need to pass legislation to combat it, as with the UK Modern Slavery Act (2015) and the US Trade Facilitation and Trade Enforcement Act of 2015, which bans the import of more than 350 goods produced by child or forced labor. Rather than a vestige of pre-capitalist social relations, unfree labor is clearly very much a part of the contemporary global economy.

The case studies above, then, detail some of the worsening labor conditions experienced by workers in the industries covered, together with how employers create labor forces in particular locations – whether FIFO workers, children on cocoa farms, or fishery workers – and then fix them in place. What these cases show is that "Old Economy" labor practices are still very much with us and, in many ways, are deepening in their severity and broadening in their geographic reach and numerical scale in the era of so-called New Economy capitalism. They are integral to the global flows of iron ore, chocolate, and seafood, together with many other commodities. These Old Economy work practices not only exist cheek-by-jowl with those of the New Economy, but they are frequently their starting point – the iron ore that makes its way into the computers and other electronics that are seen as emblematic of the New Economy, for instance, is dug out of the ground under what are often life-threatening conditions, while the cocoa that may be manufactured into chocolate in modern factories in Europe originates under conditions that are often little better than a reign of terror. In this regard, the purported bright line between the world of the Old Economy and that of the New Economy seems to be largely a rhetorical fiction.

CHAPTER SIX

Meet the New Economy – Same as the Old Economy?

The New Economy was supposed to be, as former British Prime Minister Tony Blair (1998: 8) put it, "radically different" from the old one, with "[s]ervice, knowledge, skills and small business ... its cornerstones [and] its most valuable assets [being] knowledge and creativity." Such was the presumed profundity of this transformation that some saw in the New Economy's emergence a divide with what had gone before that was every bit as acute as had been the division between the industrial mass production that developed in the early twentieth century and the nineteenth-century craft work that it replaced (Piore and Sabel 1984). Much of this New Economy talk has gone hand-in-hand with the euphoric language used to describe globalization (see Chapter 3), which Japanese management guru Kenichi Ohmae (2005: 28) pronounced "a variety of plant not previously grown, belonging to a totally novel and unknown genus and species," one that is bringing about a "global economy ... powered by technology" (2005: xxvi) that is "one grand continent of opportunity" (2005: 81). New Economy work, then, was expected to be less tiresome and more rewarding than Old Economy work, as the menial jobs of the industrial age were replaced by technology. The result would be a more humane workplace and social order, with workers liberated from the grind of long work hours to live in a world of expansive leisure time and nearly boundless personal freedom.[1] Indeed, for some the New Economy would augur the "end of work," as computerization would eliminate monotonous work to leave only those jobs requiring brain power and imagination, though what people without jobs would do under these circumstances tends to be glossed over. This latter is particularly problematic for, as Frey and Osborne (2013) have advised, computerization could eliminate half of all jobs in the United States by 2030, while, as we saw in Chapter 5, technology does not appear to be improving the lives of workers who will be laid off in places like the Pilbara, as its mines move towards workerless operations. However, although many have argued that the New Economy's rise represents, at least for some people, an emancipatory development, the situation for hundreds of millions of others is little changed from the heyday of the Old Economy. Indeed, as we saw in Chapter 5, in many ways it is actually worse. This is even the case for many of those who labor in the so-called New Economy sectors of the information age and personal services, where millions continue to toil under Taylorist, almost slave-like conditions. Moreover, even the vaunted knowledge jobs of the New Economy are frequently dependent upon workers engaged in drudge jobs that are organized little differently from how they were before the purported New Economy's birth.

Given the above, this chapter explores the underside of the New Economy – that is to say, the jobs necessary to make it function. Specifically, I focus upon three groups of workers whose labor is essential for ensuring that New Economy jobs can be done: workers who manufacture some of the technology required to make the New Economy function; workers who labor in the call centers of places like India that serve as the hubs for much of the commercial information that flows across the planet, and whose work connects consumers and producers globally as part of Ohmae's "one grand continent of opportunity"; and the personal care and janitorial workers whose labor is necessary to reproduce the workplaces and workers at the center of New Economy narratives. As I show below, then, New Economy jobs are supported by an array of Old Economy labor practices.

Laboring in the New Economy

Chips off the old (economy) block?
Arguably, if any individual piece of personal technology has come to embody the new knowledge economy, it is the smartphone. Through something the size of a pack of playing cards, any human who possesses such a device can access vast oceans of knowledge and conduct innumerable activities with a flick of the finger – read email, pay bills, map out a travel route, take a photograph and send it to a friend, post a video to the Internet, watch a movie, or even find a romantic partner. The smartphone is emblematic of modern capitalist society for another reason, however: its constant upgrading (Apple updates its iPhone about every twelve months or so). Indeed, the knowledge economy of contemporary capitalism is dependent upon the planned obsolescence of such technologies and the computer chips which allow them to function, a reality readily acknowledged by at least one key actor in the emergence of the knowledge economy – as Andrew Grove, former CEO of Intel, once remarked concerning the launch of a new chip: "This is what we do. We eat our own children, and we do it faster and faster ... that's how we keep our lead" (quoted in Ramstad 1994). The New Economy, then, is underpinned by the manufacture of computer chips and other hardware that makes the production and manipulation of knowledge possible. Significantly for the argument that we have entered an era of fundamentally new ways of organizing

work that will liberate workers, however, these material artifacts of the New Economy are generally produced in ways that are redolent of the nineteenth-century factory.

A close examination of how smartphones are produced shows that their manufacture is built upon a host of highly exploitative labor practices. Even before the components from which they are assembled are brought together in a factory, the conditions under which their constituent raw materials are extracted from the earth are hardly emblematic of supposed New Economy knowledge work. For instance, one of the key components that goes into smartphones, computers, DVD players, and many other devices of the modern age is tantalum, a highly corrosion-resistant rare metal mined in Brazil, Australia, China, and several parts of Africa, including Rwanda, Mozambique, and the Democratic Republic of the Congo (DRC). In Africa, miners, leading hardscrabble lives, and using bare hands and shovels, typically dig the mineral out of the ground in the form of coltan, a mixture of columbite and tantalite (Nest 2011). In the DRC, ongoing civil war has led various war lords to ramp up production so as to earn income, a practice that has resulted in children – some as young as 6 years of age – making up a significant proportion of the mining workforce, and local villagers suffering at the hands of rebels who frequently engage in mass rape of women and the murder of any miners who resist them. While the way in which the coltan is dug out of the ground often results in landslides, which kill miners, initial processing involves its being crushed to remove impurities like iron, a process that exposes workers' lungs to large quantities of dust. Once extracted, the tantalum is then processed in refineries in places like China, Afghanistan, and the United States before being sold to electronics manufacturers, often through shadowy networks of dealers and intermediaries, which makes tracing its origins – especially whether or not it has come from conflict zones – very difficult. Although several manufacturers have sought to protect themselves from negative publicity by announcing that they will not use conflict minerals, in practice this is easier said than done, especially if they use recycled tantalum that has come from e-waste processed in some of the global destruction networks described in Chapter 3. Other minerals that also go into smartphones, such as gold, tin, and tungsten, are frequently dug out of the ground in similar fashion.

If securing some of the raw materials that are used in making smartphones is one aspect of the highly exploitative labor practices used in bringing them to market, their actual assembly is another. Major manufacturers like Apple and Samsung have been criticized for the conditions under which their employees produce the millions of smartphones that help the contemporary global economy to function. In the case of Apple, the Foxconn plant in the Longhua district of Shenzhen in southern China, to which Apple subcontracts manufacture of its iPhones and iPads, has been the object of global disapprobation, not least for the fact that its working conditions are so bad that the company has had to put up nets to stop workers from jumping off the factory's roof and has forced workers to sign documents stating that their families will not sue the company should a worker commit suicide while employed at the plant. Foxconn itself is a Taiwanese company that is the largest private employer in mainland China (more than 1 million workers). It manufactures electronics for some of the world's best-known brands, including BlackBerry, Cisco, Hewlett-Packard, Dell, Motorola, Nokia, Nintendo, Sony, and Toshiba. Between one-quarter and half a million workers (depending upon the source) work in a walled area of Longhua (which is sometimes called "Foxconn

City"). Many live in company-controlled dormitories and rarely leave the facility, as they work up to twelve hours a day, six days a week. Workers living in these dormitories are usually placed with roommates who work different shifts and who speak different Chinese dialects, as a way to limit the development of worker resistance to the factory's managers. In one damning report, Foxconn's facility was described as a "labor camp" in which workers were often forced to work up to one hundred hours of overtime per month (sometimes without pay), nearly three times the legal limit under China's labor law (*South China Morning Post* 2010). Workers who engage in disruptive behavior are often forced to write confession notes, which are then posted in the factory as a way to humiliate them.

Another major technology company whose products have helped fuel the growth of the knowledge economy is South Korean giant Samsung. Like Apple, Samsung has faced significant criticism for how it treats its employees. In its 2016 report *Samsung – Modern Tech, Medieval Conditions*, the International Trade Union Confederation and IndustriALL, one of the global union federations (GUFs) that represents workers in particular industries (see Chapter 7), detailed how Samsung workers have been exposed to many of the same kinds of abuses and exploitative practices as have Apple's. For instance, there have been more than 200 cases of serious illnesses documented among young, former Samsung assembly workers, including several forms of cancer. In Brazil, the government filed a lawsuit in 2013 alleging that Samsung had imposed dangerous working conditions on 6,000 employees at one facility, and even went so far as to issue an arrest warrant for Samsung chairman Lee Kun-hee. In Europe, Samsung has been investigated for employing Hungarian electronics workers at wages insufficient to support a family. It has likewise been accused of underpaying apprentice workers at its Noida research and development facility in India and of union busting in Indonesia, the Philippines, and Thailand. This anti-union policy has an effect across the whole Asian electronics industry, as many subcontractors resist – often violently – unions for fear of losing their contracts with Samsung. In Mexico, Samsung has subjected women to pregnancy tests and denied them employment if the tests come back positive. In Taiwan and Vietnam, the company has been condemned for making its workers labor extremely long hours for meager pay – interviews with Samsung workers in Vietnam have indicated that forced overtime is a standard part of the work regime, to the extent that managers stand in the doorway preventing workers from leaving at the end of their shifts. According to China Labor Watch (2012), employees at Samsung's Chinese factories, some of whom are knowingly hired with false documents indicating that they are above the legal minimum working age of 16, are forced to work dozens of hours of overtime every month, standing for eleven to twelve hours a day while suffering significant verbal and physical abuse. In one court case filed in the United States, a Samsung designer testified that, because the company was rushing its Galaxy tablet to market, she was forced to stop breastfeeding her small infant so as to keep up with the roll-out schedule (*San Francisco Chronicle* 2012). Meanwhile, a leaked Samsung document has revealed how managers are encouraged to "dominate employees" through means that will "isolate employees," "punish leaders," and "induce internal conflicts," while the Asia Monitor Resource Centre has reported instances in which Samsung corporate representatives have tapped workers' phones and threatened their families.

Not only are the conditions under which smartphones are assembled often fairly poor, but so too are the situations faced by the millions of workers who manufacture the chips upon which much of the information revolution relies. The vast majority of these chips are produced in the Global South, typically in export processing zones (EPZs) where labor and environmental safeguards are fairly weak and global corporations are lightly taxed. One of these, established in 2000 in Chengdu, central China, has become a global hub for the high-tech industry, with Intel, the world's largest chip maker, opening an assembly and testing facility there in 2003 and investing more than US$600 million since then. Partly as a result of such investment, by 2013 over 20% of the computers sold worldwide were being manufactured in Chengdu, while more than half of all the chips used globally in laptops were being assembled and tested there (*Shanghai Daily* 2013). In the case of Mexico, whose border with the United States is the longest between a Global South and Global North country anywhere on the planet, such zones were first established in the mid-1960s, after the Bracero Program – begun in 1942 to bring Mexican workers to the United States to work the fields and orchards of places like California (Mitchell 2012) – was unilaterally terminated by the US government, and Mexico was suddenly faced with trying to create jobs for thousands of former *braceros*.[2] While most of these *braceros* were men, significantly the bulk of workers in the factories (termed *maquiladoras*) established in EPZs close to the US border after the program's end were initially women, as employers saw them as easier to intimidate, although in later years men were also recruited, especially in auto parts facilities.

Across Mexico, 2.7 million people today work in *maquiladoras*. Most electronics manufacturing takes place in them. Approximately one-third of this relates to the IT sector, as plants produce the computers, CPUs, circuit boards, LCD panels, network switches, routers, and other equipment necessary for the knowledge economy to function. Another third of production is of consumer electronics, like TV sets, mobile phones, and audio and video goods (NAPS 2016). For its part, Baja California, especially the city of Tijuana, located 500 miles from Silicon Valley, has developed one of Mexico's largest clusters of electronics manufacturers and assemblers. Its more than 120 companies, including firms with household names like Panasonic and Samsung, employ about 50,000 workers (Tijuana EDC n.d.). According to Mexico's National Institute of Statistics and Geography, overall Tijuana has more than 600 *maquiladora* plants (more than any other Mexican city) employing more than 200,000 people manufacturing everything from electronics to automobile parts to medical devices to components for the aerospace industry. Typically, however, wages are extremely low in these plants, largely because independent labor unions are severely repressed. In 2015, for instance, line operators at Foxconn's plant in Ciudad Juárez were making around 650 pesos a week (about US$39), while purchasing enough food for a family with children required 700–800 pesos a week (Bacon 2015). Such low wages are not confined to Mexican *maquiladoras*, though. Even within California's Silicon Valley, the community that has arguably been most associated in the popular imagination with the knowledge economy, the kinds of neoliberal policies described in Chapter 4 are impacting workers and undermine the starry-eyed image of "New Economy" working conditions – there is significant use of temporary and subcontracted labor, together with high levels of home-based work, producing parts for Silicon Valley's electronics industry (Carnoy et al. 1997; Benner 2002).

In addition to issues of work precarity faced by many of the workers who produce the technology upon which

the New Economy is based, there is also a significant disjuncture between how the electronics industry frequently
presents itself as "green" (e.g., see Apple Inc. 2016) – a
greenness in line with the expected environment-friendly
economic model upon which the future will supposedly
be based – and its actual production processes. Whereas
manufacturing computer chips seems to epitomize in
many ways the "clean" industries of the New Economy,
with workers wearing polyester "bunny suits" and gloves
and goggles while working in "clean rooms" where spot-
lessness is a requirement for producing such chips, in
actual fact the elaborate techniques undertaken to ensure
cleanliness in this most modern of workspaces are not for
the benefit of the workers but to ensure that the chips that
they fabricate are not damaged in any way. While robots
do much of the work, highly toxic chemicals like arsenic,
toluene, cadmium, trichloroethylene, and methyl chlo-
ride are widely used in the production process. Although
industry representatives typically claim that workers who
handle such chemicals face no dangers, there have been
many reports of them suffering health conditions related
to their work – an early study reported that women working in the chip fabricating areas of fourteen manufacturers
were 40% more likely to suffer a miscarriage than were
women who worked in other parts of the semiconductor
industry, with the miscarriages largely linked to exposure
to ethylene-based glycol ethers (*Los Angeles Times* 1992).
Other studies have reported similar issues, with workers experiencing higher levels of cancers associated with
exposure to specific chemicals used in the industry (*New
York Times* 1996). The impact of the industry's labor conditions upon its workers has been implicitly recognized
in a number of legal settlements. Hence, in 2003 the IBM
Corporation made confidential financial settlements with

several former employees of its San José, California, chip manufacturing plant, while in 2014 Samsung issued an apology and offered to pay compensation to several workers and their families who had suffered and/or died from leukemia, lymphoma, brain cancer, and other serious diseases, although the company did not go as far as admitting responsibility for these ailments (*International Business Times* 2014). Workers in this New Economy sector, then, face both employment and biological precarity, as toiling in this industry can dramatically shorten their lives.

Call centers: dark satanic mills of the New Economy?
Whether helping customers with a software problem or an overcharge on a credit card bill, or returning a sweater ordered in the wrong size, in many ways call centers are emblematic of the New Economy, serving as what *Fortune* magazine (1996) once called "meta-department[s] of consumer intelligence" because they break down "the traditional barrier between customer and company, [thereby] fostering ... 'customer intimacy'." In this regard, call centers have become a source of significant analytical interest, for they serve as important control points in the global economy through which flows much of the information that connects consumers in countries like the United Kingdom or the United States with representatives of the companies with which they have business relationships. Call centers, however, have also become objects for mockery because, with their legions of cubicled workers tethered to telephones working like drones around the clock, they are viewed in many ways as being exactly opposite to the liberated workplace of the imagined New Economy. In the 2006 comedy *Big Nothing*, for example, the character Charlie – played by American actor David Schwimmer – is placed in the "Jennifers and Stephens" section of a call

center because, according to his boss, "callers like to think they get the same service rep." Meanwhile, an Indian call center was a key location in the 2008 Oscar-winning movie *Slumdog Millionaire*, while the BBC reality TV series *The Call Centre* followed the goings-on at a Welsh center.

Although the earliest instances of firms using telephones to cold call potential customers date to the early twentieth century, and while housewives in North America and elsewhere were employed in the 1950s to call their friends and neighbors to sell goods, what we would recognize as a call center today has its origins in the mid-1960s. At that time, the development of private automated business exchanges (PABXs) allowed companies to manage large numbers of customer contacts using a single telephone number. Some analysts suggest that the British newspaper the *Birmingham Press and Mail* established the first bona fide call center in 1965, but the development by the US firm Rockwell in 1973 of the automated call distributor, which could automatically filter calls and forward them to the next available agent, revolutionized the industry by, essentially, replacing the human telephone operator with a more reliable and faster automated system. The first company that is normally given credit for making widespread use of this new technology is Continental Airlines, which used it to develop a telephone booking service. At about the same time, credit card companies began using such systems to help customers who had queries about their accounts; by 1972, for instance, Barclaycard had established a call center in Northampton (central England), and in the same year British Gas opened what may have been the world's first bilingual call center (English and Welsh) in Wales, handling 20,000 calls a week. Initially, such centers were used to handle customer queries and complaints; over time they shifted to direct-marketing activities. Indeed, as technology developed, growing numbers of firms were able to change their entire sales operations over to doing business over the phone; in 1985, Direct Line, which was the Royal Bank of Scotland Group's insurance division, became the first British insurance company to operate entirely over the phone. The cheapening of computer technology and changing consumer attitudes about doing business over the telephone meant that, whereas in the 1960s and 1970s it tended to be only large TNCs that operated in this manner, by the 1980s smaller firms were also getting in on the act. Indeed, these developments allowed many to outsource much of this activity to independent companies who would make calls on their behalf. Thus was born the modern call center, with the first published use of this term being in 1983.

Call centers have become major employers in many Global North countries. In the United States, Pittsburgh, which lost thousands of steel jobs in the 1980s, had become a center for the industry by the 1990s (Benner 2007). States like Utah, Nebraska, West Virginia, Iowa, and South and North Dakota also became homes to call centers, largely because they offered cheap labor. However, as the industry matured, it began restructuring geographically, such that, by 2012, Texas, Florida, Ohio, Arizona, and Colorado had become the leading states for call center employment. Significantly, with about 8,500 jobs, Idaho had the nation's highest per capita call center employment (about four times the US average) but also its largest share of minimum-wage workers. Although these two figures are not necessarily connected, it is true that call center work is typically low-waged work. In 2012, of the estimated 453,000 US call center workers, the two major occupations were customer services representatives (42% of employment) and telemarketers (29%), which paid a median hourly wage of US$11.96 and US$9.98 respectively (US Bureau of Labor Statistics 2012). Assuming a work year of 2,000 hours, this resulted in annual earnings of US$23,920 and US$19,960 (in 2012, federal government-determined poverty level wages for a family of four were US$23,050). For its part, in the United Kingdom, old industrial cities like Belfast, Manchester, and Birmingham have also seen call centers open, although they have also sprung up elsewhere – some have preferred places like Oxford and Cambridge (where lots of students look for part-time jobs), London (lots of immigrant workers), or more rural areas (like the Isle of Lewis, off the Scottish coast), where workers have few other employment options.[3] These, too, are typically low-waged work – in 2014, a call center worker's average salary was £15,206, compared to the national average of £27,271 (UK Office for National Statistics 2014).

During the 1990s, many European companies began establishing pan-European, multilingual call centers in cities like Dublin and Amsterdam. Increasingly, this ability to offer services in several languages has shaped the European industry. Significantly, though, it has often relied upon the language skills of immigrants. For instance, in 2009 TeleTech UK (a subsidiary of US firm TeleTech Holdings) announced that it would expand its center in Belfast, Northern Ireland, by hiring native speakers of Czech, Danish, Dutch, Finnish, French, German, Hungarian, Italian, Norwegian, Polish, Portuguese, Russian, Spanish, Swedish, and Turkish (*Irish Times* 2009). Although the industry remains concentrated in northern Europe (see Table 6.1), in recent years Eastern and Southern Europe have emerged as booming focal points, largely because of cheaper labor costs and language skills. Romania, for instance, has some 30,000 employees working in call centers or in the general business process outsourcing (BPO) sector, with about half of

Table 6.1 Top ten European countries for call center employment, 2012		
	Employed	*% of total*
United Kingdom	1,041,500	31.0
Germany	450,000	13.5
France	273,000	8.1
Netherlands	220,130	6.5
Italy	200,000	5.9
Spain	173,000	5.1
Poland	143,900	4.3
Sweden	102,421	3.0
Russia	100,000	3.0
Belgium	87,500	2.6

Source: ECCCO 2012: 5

these in Bucharest. Even here, though, the urge to reduce labor costs ever further is always present. In 2012, TELUS International Europe (formerly CallPoint), founded in 2007 to provide services in Romanian, English, French, Italian, German, and Spanish to clients in Western Europe and North America, decided to open a new center away from its main base of operations in Bucharest because the competition there for workers had become too great; the company set up shop in the smaller city of Craiova. Meanwhile, in the 1990s many US firms began outsourcing operations to Canada and India – whereas India offered the advantages of low wages and a well-educated, English-speaking population, Canada's political and economic stability, proximity, and cultural similarity to the United States meant that it tended to attract higher-value, more sophisticated work. Canadians' greater awareness of US fashion trends also meant that it was preferred over India by many mail-order fashion retailers (*New York Times* 2004).

In discussing the geography of call center locations, though, arguably it is the growth of centers in India that has most grabbed the public imagination. Indeed, from 2010 to 2011, the US television channel NBC even had a sitcom (*Outsourced*) set in a Mumbai call center, with the series based upon the 2006 romantic comedy film of the same name. The industry's origins largely lie in decisions taken in the mid-1990s by US and British companies like GE Capital, American Express, Aviva Global Services, and Barclays to situate back-offices in India to take advantage of large numbers of English-speaking graduates who had IT and other skills – much of the early industry developed in places like Bangalore, India's Silicon Valley. Many Indian centers even recruited entry- and mid-level employees from European countries (on Indian salaries) to service European customers, with such workers liking the notion of living an "Indian adventure" (*BBC News* 2004). By the mid-2000s, India accounted for 46% of all offshored BPO workers globally (more than 400,000 workers), with about 65% of these employed in call centers (NASSCOM-McKinsey 2005: 13; Taylor and Bain 2006: 19 and 21). However, by the end of the twenty-first century's first decade, many firms had begun to look elsewhere, the result of a lack of sufficient numbers of skilled workers, poor quality urban infrastructure, complaints from some customers that Indian workers' accents were too thick for West European and North American ears, and a lack of workers fluent in French, German, Japanese, and Spanish. The shortage of language skills has made several other Asian and East European countries more attractive to Japanese and West European companies. Many US firms have now opened operations in the Philippines, a former colony with close cultural ties to the United States and a population of educated workers fluent in English and Spanish. Although wages tend to be higher than in India, overall costs are generally lower (training is less expensive because Filipino workers tend to have higher levels of formal education) and more languages are typically offered – Filipino centers offer services not only in the industry's traditional European languages, but also in Mandarin, Cantonese, Taiwanese, Korean, Bahasa Indonesia, Bahasa Malaysia, and Thai (*Infinit Contact* 2014). Consequently, growing numbers of firms have shifted their operations there – between 2011 and 2013, some 75,000 BPO jobs relocated from India to the Philippines (Arun 2013), where, in 2016, about 1.2 million call center workers were employed (*Wall Street Journal* 2016). Meanwhile, rising costs and a lack of language skills among Indian workers has allowed European countries like Portugal to benefit from the growing trend of "near-shoring," wherein increasing numbers of European firms are basing their call centers closer to home rather than in Asia – among other things, doing so means that issues of currency conversion and operating across numerous time zones are of much less concern. Some firms have even relocated back to the United States, taking advantages of tax breaks and other financial incentives offered by various state and local governments – states like Ohio, whose population tends to speak with an accent not particularly identifiable with any given region, have seen companies like General Electric, Amazon.com, and Barclays Services LLC open up facilities in recent years.

The growth of call centers in countries like India, the Philippines, and Romania has undoubtedly provided jobs and incomes to workers who otherwise might not have had them. In the Global South, many of these workers are women, who are often seen by managers to be harder working and more patient and friendly than men. Call center work, then, has given hundreds of thousands of women opportunities for financial independence that they otherwise would not have had and so has allowed them to challenge many of the norms of the patriarchal societies in which they live. However, patriarchal attitudes and practices do persist. For instance, many parents from more conservative rural areas have concerns about their daughters working night shifts with men and how this may lead to sexual encounters. Likewise, research suggests that many of the gendered divisions of labor that existed in the Old Economy have continued into the New Economy, as men appropriate higher-skilled and higher-paying jobs and women are consigned to lower-skilled and lower-paying ones. Studies on the household division of labor among Indian call center workers have also shown that men tend to use technology to continue working at home, often in a separate room, while women workers are more likely to find themselves having to combine telecommuted call center work with domestic responsibilities – that is to say, traditional gender roles are maintained as women must fit their paid work around their household chores (Pradhan and Abraham 2005). Some commentators have suggested that call centers reduce young workers to being little better than "cyber coolies" who work under sweatshop conditions, "except that you're working with computers and electronic equipment rather than looms" (Bidwai 2003). Call center workers also report high levels of stress caused by pressure to meet sales targets and habitually having to work throughout the night – time differences mean that a customer calling from the United States in the late afternoon might be speaking with a worker in Bangalore or Manila for whom it is the middle of the night, with the result that many workers find themselves becoming disconnected from circles of friends and families as their own bodies' circadian rhythms must adjust to those of their customers in a kind of temporal colonialism.[4] Additionally, many workers feel cut off from their local communities and experience high levels of emotional exhaustion because they are often required to engage in a degree of geographical subterfuge by assuming fake identities and speaking with British or US accents, depending upon where the center's client is located (Poster 2007). To relieve their stress, many engage in risky sexual behavior and/or excessive drug and alcohol use (Vaid 2009; Charles et al. 2013). The nature of the work – sitting for long periods – means that many also suffer from high levels of musculoskeletal and gastrointestinal disorders, as well as headaches/migraines (Upadhyay et al. 2012).

Although call center workers' conditions of labor, then, often do not conform to the popular image of self-actualizing knowledge work, such conditions vary considerably, depending upon where call centers are located. In "social market" economies with relatively strong labor market regulations and relatively influential labor market institutions (like Austria, Denmark, and Germany), job quality tends to be better, there is lower employee turnover, and wages tend to be higher. In more liberal market economies, with more relaxed labor market regulations and less influential labor unions (like Canada, Ireland, the United Kingdom, and the United States), and in recently industrialized or transitional economies (like Brazil, India, Poland, South Africa, and South Korea), however, work is often more repetitive and call centers tend to make greater use of contingent labor, though this varies by country – the Indian labor force has traditionally largely been full time, for example (Holman et al. 2007). There are, though, great changes ahead for the industry, changes that are likely to make conditions worse for millions of workers, for just as technology is impacting many other aspects of workers' lives, so is it affecting call center operations. To cut labor costs, many centers are replacing voice work with email and encouraging their customers to use chatbots to address basic service questions. Meanwhile, other technologies (such as British company Celatron's inStream system) are being developed that will refer certain tasks to human operators and remember their responses, with the goal ultimately of not needing such humans – as the company puts it: "With each transaction, inStream is continuously learning, reducing the need for human intervention in the future" (Gidda 2016). With the automation of much of this low-end work, the industry has increasingly turned to more value-added "non-voice" activities, like high-end analytics and cloud computing. Although this potentially provides some opportunities for higher pay for workers with such skills, for hundreds of thousands of call center workers across the globe it is likely to result in job losses. As such basic jobs disappear, those workers who are left will be needed not for repetitive tasks but, rather, to sell customers various goods and services, a task requiring better language skills that will likely lead some of the jobs presently being done in places like India and the Philippines to be brought back to Europe and North America (*The Economist* 2016).

Ghost workers of the New Economy
One popular New Economy image is that of the spotless workplace, wherein creative types lounge around as they work. Combatting dirt is viewed by many as central to reproducing the modern workplace as a healthy one for workers. Clean work and clean workplaces, it is regularly suggested, go hand-in-hand in the New Economy. At the same time, ensuring the social and biological reproduction of the New Economy's labor force on a daily or generational basis is also central to its success. Significantly, though, much of

this reproduction work is done on the backs of millions of domestic workers – entrepreneurs at a Silicon Valley tech firm are often only able to be so deeply embedded in the work that characterizes the New Economy because they rely upon substantial help from nannies, housekeepers, gardeners, and a host of other people who raise their children and keep their household functioning. In many ways, however, cleaners and domestic workers are the ghost workers of the global economy. Often, they labor when most of us are sleeping or are outside the home and we only recognize their job when it has not been done – frequently, it is the absence of a clean workplace or a cooked meal or washed clothes that we notice, not their presence. Although it is often invisible, the unglamorous labor of janitors and domestic workers, then, is central to reproducing the workplaces and workers of the New Economy.

While Hollywood has romanticized the life of cleaners in movies like *Maid in Manhattan* and *Good Will Hunting*, such work is actually one of the most injury-prone activities in the New Economy. In 2015 in the United States, for instance, the rate of injuries and illnesses with days away from work for janitors and cleaners was 657.4 cases per 10,000 full-time workers, a rate higher than for firefighters, police officers, correctional officers, truck drivers, and laborers. Most of these injuries involved cuts, sprains, strains, and fractures, most of which were sustained by overexertion in lifting, falls and slips, exposure to harmful substances, and being hit by objects. Janitors and cleaners also suffered much higher rates of musculoskeletal disorders than the average worker (US Bureau of Labor Statistics 2016). For this, they received a median annual income of only US$23,440, compared to that for all workers of approximately US$36,200 (US Bureau of Labor Statistics 2015b).

Similar trends can be found in other countries. In 2010 in Australia, the federal government set the national wage rate for the industry at AU$18.46, barely above the national minimum wage. However, the Fair Work Ombudsman (2011) found in an audit of some 315 cleaning companies that nearly 40% were paying below this rate or otherwise violating the law, and that cases of unfair dismissal, underpayment, and sham contracting, in which employers misrepresented employees as independent contractors to avoid paying standard entitlements, were common. If a similar rate pertained across the whole industry, that would mean that about 6,000 of the 15,000 cleaning firms on the Australian Business Register at the time would likely have been in some violation of the law. Follow-up investigations in 2013 and 2015 found similar violation rates (Fair Work Ombudsman 2016). One reason for this is that many cleaners are international students who tend not to know their legal rights; according to a 2009 survey by United Voice, the cleaning sector union, 58% of respondents were international students (*The Age* 2014). In Europe, although a paucity of official statistics makes it difficult to determine exact numbers of work-related health problems, data show that in Germany cleaners are off sick an average of 24–27 days annually (50% higher than the national average), while in Sweden it is 13 days (the third-highest rate of all occupations), with female cleaners having twice as many reported work-related diseases as the average for all employed women (SCA 2012). In a 2008 survey of interior cleaners, 74% reported experiencing muscular aches, pains, and discomfort in the previous year (EU-OSHA 2008). Wages are also low, undoubtedly a reflection of the fact that the industry's labor force is predominantly female, part time, and a large proportion are immigrants (EU-OSHA 2009).

The cleaning industry is highly labor-intensive – about 80% of costs are for labor – and so is under constant pressure to keep wages low. Doing so has been facilitated by the increasing contracting out of janitorial services to independent cleaning firms that bid on the work, which has allowed workplace managers to play these firms against each other for the lowest price and also to divest themselves of responsibility for the healthcare and long-term pension costs of those who clean their workplaces. Outsourcing to keep labor costs down has been complemented by growing pressures upon cleaners to increase productivity. Aguiar (2001), for instance, notes that adoption of total quality management techniques in the Canadian industry, whereby workers are subjected to time and motion studies to measure the "proper" amount of time that should be taken by any particular task, has been used to reduce the number of cleaners assigned to particular job sites. This has not only intensified the labor process, but has also reduced janitors' autonomy to use their own skill to determine what constitutes a "clean" workplace. Combined with the loss of much of the previously enjoyed social safety net brought about by neoliberal-inspired labor market deregulation (see Chapter 4), the disintegration of many workplace rights and ever greater pressures to cut corners has led to the emergence of what Aguiar (2006) has termed "sweatshop citizenship" for thousands of cleaners. Such developments have consequences, though, not just for the janitors themselves but also for society more broadly – at least one study from the United Kingdom has suggested that the outsourcing of cleaning has led to a reduction in standards, which has played a role in the spread of various infections in hospitals, including the virulent methicillin-resistant *Staphylococcus aureus* (MRSA) bacterium (Davies 2005).

Work conditions for many cleaners have also been transformed by significant structural changes in patterns of ownership in the cleaning industry. Although the vast majority of cleaning firms are relatively small, there have emerged some true giants, including Danish firm ISS (more than 460,000 employees worldwide in some fifty countries), ABM Industries Incorporated (based in San Francisco, with more than 100,000 employees in twenty international locations), and GDI (the largest facility services provider in Canada and one of the five largest in North America). Although, to date, the commercial cleaning industry has largely been concentrated in Global North countries like Australia, Canada, Japan, the United States, and Western Europe, demand has been growing in countries like China and India, which has encouraged some of these cleaning TNCs to establish operations there. The growth of these larger firms means that, whereas some workers have managed to secure permanent work in the same locations because the cleaning firms have used their size to negotiate long-term arrangements with the buildings' owners, the fact that decisions over conditions and wages are usually made at some considerable distance from the job site – perhaps even in another continent – means that these cleaners have little influence on decisions regarding their work life and compensation. Finally, growing consumer demands for "green cleaning" have also been affecting the industry, especially concerning the types of solvents used, although the impetus for this has generally been more out of concern for the environment than for cleaners themselves.

If commercial cleaners comprise one group that is central to the reproduction of the New Economy workplace, the coterie of domestic workers who help many of its knowledge workers is another. Although it is difficult to determine exact numbers, the ILO (2011) estimates that there are, at a minimum, 50 million domestic workers globally, with Asia and the Pacific and Latin America and the Caribbean having the greatest numbers of such workers (41% and 37% respectively of the global total) (ILO 2013b: 24). Not all of these work for the middle-class knowledge producers and managers who are imagined to be at the heart of the New Economy. But a fair few do. In Britain, for example, although accurate figures are hard to come by, it is reckoned that in the early twenty-first century there were about 2 million domestic workers, more than during the Victorian era, when, it is often imagined, the very embodiment of class hierarchy involved having servants in the home (Cox 2006). In the United States, the Obama administration's Secretary of Labor Hilda Solis (2012) suggested that in 2012 there were about 1.8 million domestic workers engaged in cleaning, childcare, and caring for the elderly and the disabled. Both of these figures are undoubtedly significant underestimates.

For every knowledge worker lauded in the press, then, there seems equally to be a domestic worker embroiled in the daily grind of household chores and/or suffering at the hands of their employer. This appears particularly so in the case of domestic workers who have come from overseas to take on childcare and housekeeping jobs, with newspapers around the world rife with stories of employers who abuse those who work for them – stories of rape or even murder are not uncommon. At the less extreme end of the spectrum, it is commonplace to find that such workers are grossly underpaid and often subject to injury as they carry out their tasks. A US survey conducted in 2012 found that 23% of domestic workers were paid below their state's minimum wage (for those who lived with the families for whom they worked, the figure was 67%), that few receive

any benefits like health insurance or paid sick days, that
25% of live-in workers had responsibilities that prevented
them from getting at least five hours of uninterrupted sleep
at night, that 38% had suffered from work-related musculoskeletal pain in the previous twelve months, and that 91%
who encountered problems with their working conditions
did not complain for fear of losing their job (Burnham
and Theodore 2012). This same survey also found wages
to be highly racialized (white women tended to earn more
than black, Asian, and Latina women) and also connected
to immigration status – undocumented workers typically
earned about 20% less per hour than did citizens. The
lower wages paid to undocumented workers was, in part,
likely related to how they find employment – while those
who are citizens may work for an agency that provides nannies or housecleaners and are therefore more likely to work
"on the books," this is rarely the case for undocumented
workers, who more typically find their jobs through various
social networks (e.g., a cleaner's employer may recommend
her to a friend, a male family member working as a gar-
dener may ask if his wife can clean his employer's house,
etc.). The fact that many such domestic workers find their
jobs through such social networks and beyond the reach of
formal labor market regulations, then, means that few are
able to set their own wages or successfully request raises,
even when workloads change (Mattingly 1999).

Although many nannies, house cleaners, and other
domestic workers are no doubt treated well by their
employers, lots do nevertheless find themselves in positions in the labor market which leave them open to
exploitation. For one thing, the fact that they tend to
work behind closed doors (whether they are citizens or
undocumented), frequently living cheek-by-jowl with their
employers, means that they are generally beyond the gaze

of the state's regulatory apparatus. This is exacerbated by the fact that in many countries they are not protected by various labor laws. In the United States, for instance, domestic workers are specifically excluded from the provisions of the 1935 National Labor Relations Act. In many other countries, they are likewise exempt from legislation on working time, minimum wages, and health and safety (Mantouvalou 2012). While some governments have attempted to address these deficiencies – in 2008 the State of New York passed a Domestic Workers' Bill of Rights, which included protection from discrimination and harassment, as well as providing for paid leave – others have moved in the opposite direction. In 2012, the UK government changed how it issues visas for immigrants hoping to work as domestic workers. Specifically, in an attempt to crack down on unskilled immigrants who were staying in the country illegally, the government switched to a policy whereby domestic workers' visas would be tied to their employers. This effectively bound these workers to specific individuals and so limited their ability to find better work. Likewise, the *kafala* system practiced in many Gulf states, whereby immigrant workers must have an in-country sponsor (typically, their employer) who is responsible for their visa and legal status, and the A3 and G5 visa system used by diplomats and employees of international organizations to bring workers into the United States as domestic employees both also bind workers to specific employers. To make matters worse, many employers confiscate workers' passports, forbid them from going outside, and do not allow them to contact their families. Certainly, there have been attempts to ameliorate the conditions under which domestic workers toil internationally. In 2011, the ILO passed its "Decent Work for Domestic Workers" convention (#C189), aimed at some of the most common exploitative practices faced by domestic workers. However, such conventions must be endorsed by individual national governments and, as of 2016, only twenty-three had done so.

One of the most significant aspects of the work of reproducing – biologically and socially – the households of New Economy workers is the fact that much of this relies upon what Hochschild has termed "global care chains," by which she means "a series of personal links between people across the globe based on the paid or unpaid work of caring" (2000: 131). Indeed, in recent decades, the growing demand for domestic workers – between 1995 and 2010, the number rose from 33.2 million to 52.6 million globally (ILO 2013b: 24) – has been one of the main drivers of the mass migration of Global South women. At least some of this migration has been boosted by reductions in social welfare spending as part of the neoliberalization of Global North economies, which has left many people with no choice but to find private solutions to solve issues of how to provide, say, healthcare to housebound family members. Globally, one in every thirteen female waged workers is a domestic worker, though in some parts of the world the proportion is higher – 26.6% in Latin America and the Caribbean and 31.8% in the Middle East (ILO 2013b: 20–21). In several regions, including Europe, the United States, the Middle East, and the Gulf states, the majority of domestic workers are immigrants (ILO 2010: 6). Historically, many female domestic workers accompanied their husbands as "dependent" or "secondary" migrants, and then sought work in the domestic arena wherever they ended up. Increasingly, however, much international domestic worker migration is being facilitated by various labor-recruitment agencies, or even by national governments themselves. In the Philippines, for example, the government has long encouraged its nationals to work overseas as a way to secure foreign exchange for the country. Whereas men have tended to go to the Gulf states and places like Taiwan to work in the oil industry or in construction, women have tended to migrate to places like Europe, Hong Kong, and the United States, seeking jobs as caregivers in private homes or in hospitals or nursing homes. Furthermore, according to Reuben Seguritan, lawyer for the Philippine Nurses Association of America, the Philippines is the world's largest supplier of trained nurses, with 429 nursing schools turning out 80,000 students a year (Rodis 2013), many of whom end up working in the United States – partly due to its colonial relationship with the Philippines (Choy 2003) – or in other English-speaking countries.

Many domestic workers who migrate from the Global South to the Global North do so because they anticipate a better life, and some undoubtedly achieve it. However, for others, the dream is often little better than a nightmare. Many women find that, despite being promised jobs as maids or nannies, they have in fact been trafficked into the sex industry. Others find themselves subjected to racial or other stereotyping. Hence, Anderson (2003: 108) recounts being told by a Ghanaian domestic worker in Athens, Greece, that her employer's daughter would not take a glass of water from her hand because of the blackness of her skin, while some families in Hong Kong have preferred to hire Indonesians or Thais rather than Filipinas as domestic workers because they considered the latter too Western and savvy and so less easy to manipulate (Constable 2007). Many migrant workers with fairly high levels of education often find themselves doing work that requires much less skill than they have because they have been categorized as immigrants and, therefore, as "inferior workers" (Pratt 1999). Equally, those who have been hired via agencies consistently find that they are almost perpetually indebted to them, as they work to pay off fees they were charged to secure their employment abroad (Parreñas 2001: 40). These workers' families are also affected by the dynamics of the global care chain; whereas, on the one hand, laboring as domestic servants in the home of a middle-class Global North employer provides a source of income to be sent back home to family members, it also means that families become transnational in structure, with mothers, fathers, or both away from their children for long periods of time. As a result, these children are often raised by relatives or locals who are hired to take the place of their parents who are working overseas, thereby reinscribing locally the practice of commercializing the biological and social reproduction of families. The global care chains that start in the homes of Silicon Valley computer programmers, financial whizzes in Hong Kong, software engineers in Britain's Silicon Fen, or IT innovators in Bangalore, in other words, have very long tentacles that can stretch into thousands of geographically isolated communities across the Global South.

Summary

Whereas the image of the New Economy put forth by many is of workers at Google or Microsoft lounging around and playing games so as to let their creative juices flow, this knowledge economy is built upon the backs of millions of poorly paid workers who are essential to its operation – workers who manufacture the hi-tech gadgets necessary for information to whiz around the globe, call center phone operators who manage much of that information, or cleaners and domestic care workers who ensure that New Economy workplaces and the homes of those who labor in them are maintained. This means two things. First, for millions of workers who face ever more precarious work and who toil under appalling conditions to make the commodities that characterize it and to facilitate its reproduction on a daily or generational basis, the New Economy does not seem to be as utopian as it is frequently presented. Second, the fact that much of the labor upon which the New Economy rests would be quite recognizable to students of nineteenth- and twentieth-century industrial capitalism, with its Fordist and Taylorist work methods, means that, rather than thinking of the Old and New Economies as being very different from each other, we must recognize that the work practices of the Old Economy remain very much embedded within the New. Rather than there being a sharp break with the kind of capitalism that came before the "radically different" New Economy that allegedly emerged in the 1990s (Blair 1998: 8), we appear to be experiencing in many ways simply more of the same. The real question to be asked, then, is who benefits from arguing that there has in fact been such a sharp break? Presumably, not the workers who find themselves embroiled within many of the labor practices of this New Economy.

CHAPTER SEVEN

Workers Fight Back

So far, I have largely looked at how workers are used as a resource by employers. In the process, I have intimated that they can shape how the economic landscape is made either by going along with or by resisting firms' decisions concerning how to deploy labor as a resource. In this chapter, I explore this issue much more fully by detailing some of the many ways in which workers have proactively created particular organizational structures and engaged in other activities to alter their own conditions of existence and so challenge how they are used by others – that is to say, I show how workers have acted as the subjects of their own histories and geographies. The chapter, then, focuses upon what labor's activism means for it as a resource. For instance, through forming labor unions, workers can increase their wages and improve the conditions under which they toil, which has significant implications for how potential employers make use of them. What is particularly vital to understand in all of this, though, is that labor's agency has not only an important history to it – workers have more power to shape the conditions of their own existence at some times and less at others – but that it also has an important geography to it. This is because workers must not only engage with the unevenly developed economic landscapes of global capitalism if they are to successfully transform the conditions under which they labor but, in so doing, they also reshape those very landscapes. In turn, this

forces employers to reconsider their own spatial organization and goals for how the economic landscape should be
constituted. As outlined in Chapter 1, the fact that working
people are sentient socio-*spatial* actors has great importance
for understanding labor as a resource and for understanding capitalism's geographically uneven development.

Workers Coming Together

One of the most powerful ways in which workers can exert
their influence is through not working – that is to say,
by withdrawing their labor. This typically has its greatest
effect when done collectively. There are many examples
throughout history of workers doing just this. Indeed,
there is evidence that some of those involved in building
tombs in ancient Egypt walked off the job when they did
not receive their food rations, while the Peasants' Revolt
in fourteenth-century England was initiated by thousands
of peasants around the country rising up in protest and
marching to London to air their grievances over, among
other things, their living conditions. However, the modern
era of workers forming organizations by which to protect
their interests really starts with the industrial revolution
in Europe. In the early days, many of these activities were
severely repressed by employers and governments. For
instance, agricultural workers from Tolpuddle in Dorset,
southern England, were transported to Australia in the
1830s to serve prison terms for their efforts to create
a union. In the United States, in 1806, Philadelphia
shoemakers were convicted of engaging in a criminal
conspiracy when they went on strike. However, as the
nineteenth century wore on, labor unions were able to
establish themselves and grow in influence. Certainly,
they are still severely repressed in many countries, but

in most democracies they are now generally accepted as legitimate representatives of workers' interests, although their organizational structures vary considerably. In countries like Germany, for instance, unions see their goals as being more than simply negotiating better wages and working conditions for their members, but also as important shapers of labor market policy, especially through the practice of "co-determination," in which union members are elected to the boards of the corporations for which they work and have voting rights to determine things like what kinds of technologies are introduced into a workplace, what kinds of products are manufactured and how, and so forth. In the United States, on the other hand, unions have more traditionally been workplace-focused and have not generally participated in determining how a firm operates its production process, or in deciding what it makes – indeed, they quite explicitly gave up much of their claim to mutually manage workplaces in the mid-twentieth century in return for regular wage increases (Harris 1982). Meanwhile, in some other parts of the world, like China, although various unions do exist on paper, these are government-run operations and typically do not serve as free expressions of workers' wills; instead, they are used largely to keep workers in line so that they do not disrupt operations in export processing zones (see Chapter 6) and elsewhere.

In general, labor unions normally organize workers according to the particular skills they possess or according to the industries within which they toil. The first of these ways of organizing workers – called "craft unionism" – was typical of much early unionism in the United States and in many European countries. Many commentators see the roots of craft unionism in the guild system of controlling work that dominated medieval Europe. For example, carpenters would come together in their union, bakers in theirs, printers in theirs, and so forth. These workers tended to be fairly skilled and formed unions as a way to prevent their privileged positions in local labor markets from being undermined by less-skilled, lower-paid workers. Whereas the first craft unions were largely focused upon working conditions in particular communities, in the middle of the nineteenth century these separate unions began to come together into larger affiliations – workers from one community would often travel to another for work and would link up with the local union there, such that over time regional and then national federations developed. This was particularly the case in industries (like engineering) in which skills are geographically transferable. In other industries, where working knowledge is much more geographically specific – as in coal mining, where geological differences between places means that familiarity with the coal seams in one region does not necessarily transfer to knowing how to work them in a different region – national union organizations also often eventually developed, though they were frequently more localist in nature, at least in Britain (Southall 1988). The second type of unions came into being in the late nineteenth and early twentieth century as many employers sought to erode craft skills by introducing machinery that deskilled workers and made them easier to replace with less-skilled labor – much of the skill of a particular activity was increasingly embodied in the machine rather than in the worker operating it. In these "industrial unions," all workers who labored in a particular industry, regardless of the job that they actually did, were eligible to join the union. In Britain, these types of unions emerged in the 1880s and 1890s when unskilled and semi-skilled workers began to join new labor organizations and some craft unions combined with industrial ones. In the United

States, they largely developed in the 1930s. In Australia, craft unions first formed in the early nineteenth century, but by the 1890s industrial unionism had also started to take hold. Typically, such industrial unions were more radical in their demands and actions than were the earlier craft unions. Both craft and industrial unions also developed various national federations: the Trades Union Congress (TUC) in Britain, the American Federation of Labor (AFL) and later the Congress of Industrial Organizations (CIO) in the United States, and the Australian Council of Trade Unions. Sometimes they were also involved in the creation of various political parties, such as the UK Labour Party and the Australian Labor Party. Through these activities, workers have shaped in noteworthy ways how labor markets operate and also influenced the broader political economy of the societies in which they live.

Significantly, workers' organizations were involved in international activities from their earliest days, thereby playing important roles in the process of globalization (see Chapter 3). As economic booms and crises began to become more synchronized in the 1840s and 1850s (Hobsbawm 1975), and as British employers in particular increasingly turned to foreign workers to break strikes, workers and their allies began coming together to form labor organizations and political societies to improve their conditions, including the International Association (founded in 1855), the Congrès Démocratique International (1862), the Association Fédérative Universelle (1863), the Ligue de la Paix et de la Liberté (1867), and the Alliance de la Démocratie Socialiste (1868) (Devreese 1988). Arguably, the most significant of these was the International Workingmen's Association (the "First International," with which was associated the radical economist Karl Marx), founded in 1864 to bring together labor activists from several countries to discuss methods for promoting international labor cooperation: one of its key goals was to secure an eight-hour work day. Many of these early labor activists were unaffiliated with formally organized labor unions. However, by the 1880s the creation of permanent national trade union centers like Britain's TUC (established in 1868) and the AFL in the United States (founded in 1886) meant that, increasingly, it was labor unions that came to speak on behalf of workers' interests, at least in the anglophonic countries. Efforts to create transnational labor links were frequently carried out through the migration of workers themselves (see Chapter 2). For instance, members of the British Amalgamated Society of Engineers and the Amalgamated Society of Carpenters and Joiners set up union branches in countries like Canada, Australia, New Zealand, South Africa, India, the United States, and elsewhere (Southall 1989). Likewise, Canadian workers linked many of their unions to US ones, like the Iron Molders' Union and the National Typographical Union, as their members crisscrossed the border in search of work, with such linkages designed to establish common employment standards across North America. Similarly, Italian, German, Portuguese, French, and Spanish unionists played important roles in the development of labor organizations in Argentina and Brazil in the late nineteenth and early twentieth centuries, with – significantly – Italians being much more influential in the creation and operation of these unions than were native-born Argentines or Brazilians (Baily 1969). Through their physical migration, then, workers transmitted ideas about unionism across space and built concrete connections between their new and old countries of residence, connections that facilitated the development of new international structures of unionism. Worker efforts to develop international labor links during this time were, in fact, so prodigious that van Holthoon and van der Linden (1988: vii) have described the century prior to World War II as the "classical age" of working-class internationalism.

Whereas the mid-nineteenth century marked a "pre-national phase" (van der Linden 1988) of transnational labor activism, as workers migrated from continent to continent and sought to build organizations that were international in scope, by the late nineteenth century trans-national labor activism had largely become the domain of the newly formed national labor entities like the TUC and the AFL and their constituent member unions. Two principal organizational structures developed. The first were entities designed to allow unions covering workers in particular sectors – coal mining, metal working, transportation, garments – to coordinate with one another across national borders. Initially called international trade secretariats, but renamed global union federations in 2002, these organizations today represent more than 100 million workers worldwide (Table 7.1). The second set of organizations were those to which various national trade union centers affiliated. In 1903, several centers met in Dublin, Ireland, to create the International Secretariat of National Trade Union Centres (later renamed the International Federation of Trade Unions). Although initially made up only of European national centers, in 1910 the AFL joined. During the twentieth century, however, the politics of the Cold War dramatically reshaped the organization, leading it to split after World War II into the International Confederation of Free Trade Unions (ICFTU), largely made-up of centers from the Western countries, and the World Federation of Trade Unions (WFTU), largely made up of communist-dominated centers. Other similar organizations also developed. In 1920, the International

Table 7.1 Global union federations

Name	Sector	Membership	Headquarters
Building and Wood Workers International (BWI)	Building, building materials, wood, forestry and allied industries	12 million in c. 350 affiliates in 135 countries	Geneva, Switzerland
Education International (EI)	Education	30 million in c. 400 affiliates in 172 countries and territories	Brussels, Belgium
IndustriALL	Mining, energy and manufacturing	> 50 million in c. 140 countries	Geneva, Switzerland
International Arts and Entertainment Alliance	Arts and entertainment	800,000 in 300 affiliates in 70 countries	Brussels, Belgium
International Federation of Journalists (IFJ)	Journalism	600,000 in 161 affiliates in 139 countries	Brussels, Belgium
International Transport Workers' Federation (ITF)	Transportation	4.5 million in c. 700 affiliates in c. 150 countries	London, UK
International Union of Food, Agricultural, Hotel, Restaurant, Catering, Tobacco and Allied Workers' Associations (IUF)	Agriculture and plantations, preparation and manufacture of food and beverages, hotels, restaurants and catering services, tobacco processing	12 million in 390 affiliates in 122 countries	Geneva, Switzerland
Public Services International (PSI)	Social services, health-care, municipal services, central government and public utilities	20 million in 669 affiliates in 154 countries	Ferney-Voltaire, France
UNI Global Union	Professional and technical employees, communications, media, and services	20 million in c. 900 affiliates in 140 countries	Nyon, Switzerland

Federation of Christian Trade Unions was founded in The Hague as a center for the Catholic trade unions that had been created in response to Pope Leo XIII's 1891 encyclical *Rerum novarum*, which addressed the conditions of workers and the poor. After working with Muslim and Buddhist unionists in Africa and Asia, in 1968 it changed its name to the World Confederation of Labour (WCL) and generally became a little more secular. With the collapse of the Soviet Union, the WFTU lost many of its East European member unions (Herod 2002), while the ICFTU and WCL merged in 2006 to form the International Trade Union Confederation (ITUC).

Both the GUFs and the ITUC play important roles in shaping how the geography of the global economy is unfolding and what this means for labor as a resource. For instance, in 2003 the International Transport Workers' Federation, founded by Dutch and British unions in 1896 to better coordinate actions, negotiated what it claims was the first ever internationally bargained, worldwide collective agreement. For its part, the Building and Wood Workers' International (BWI) GUF has been heavily involved in fighting some of the child labor abuses detailed in Chapter 5. After a lobbying campaign, in 2010 the World Bank agreed to include as part of its standard contract with financial institutions to whom it lends money for major construction works across the globe the requirement that any contractors working on these projects must respect core ILO standards on child and forced labor, gender discrimination in employment, and permitting union activities (ITUC 2011: 14–15). In India, Nepal, and Bangladesh, the BWI has worked with its affiliates to promote its Children Should Learn Not Earn campaign, which has resulted in thousands of children who previously worked in brick kilns, stone quarries, and forestry being able to go to school – as part of this, the BWI has established a network of schools for child laborers and has developed worksite and mobile schooling initiatives. The BWI has also launched global campaigns against occupational diseases and has been successful in pressuring several Latin American countries to implement bans on the use of asbestos. Meanwhile, the former International Metalworkers' Federation (now merged with several other GUFs to form IndustriALL) played important roles in the transition in Eastern Europe in the 1990s, working with newly democratic unions on matters like coordinated bargaining with West European unions that represented workers employed by some of the firms that were expanding into the East, developing new labor policies in conjunction with national governments, and helping workers in the transition from state socialism to the free market in an era of neoliberal predatory capitalism (see Chapter 4) (Herod 1998). Other GUFs have similarly been involved in coordinating globally activities involving unions and workers in particular economic sectors.

For its part, made up of more than 300 affiliates representing 176 million workers in 162 countries and territories, the ITUC has been heavily involved in promoting transnational labor solidarity, the adoption of various ILO conventions to improve workers' working and living standards, and facilitating union organizing across the planet. It claims to be the "global voice of the world's working people" (ITUC 2016). One of its ongoing campaigns has been to reduce and/or eliminate precarious work that has come to dominate many sectors of the economy. In this regard, it has challenged the *kafala* system of work sponsorship practiced in many Gulf states (see Chapter 6), focusing particularly on the labor conditions affecting workers in Qatar – the vast majority of whom are immigrant workers who are forced to live in appalling conditions in labor camps – who are building the stadiums that will be used for the 2022 soccer World Cup (ITUC 2014). The ITUC has also fought against precarious work in the Asian seafood industry (see Chapter 5) and in the global apparel industry. In seeking to foster transnational labor solidarity and cooperation, the ITUC frequently works through its regional organizations in various parts of the globe, including the Trade Union Confederation of the Americas, the African Regional Organisation of the ITUC, the European Trade Union Confederation, and the ITUC Asian and Pacific Regional Organisation. The ITUC also works closely with the GUFs to pursue various International Framework Agreements, which have been negotiated with some TNCs to ensure that labor conditions generally conform to internationally accepted standards (as of 2016, more than 90 of these agreements had been signed). Significantly, although such agreements do not provide for direct negotiations over wages and conditions, they do provide a context within which dialogue between employers and their workers can occur (Thomas 2016). Moreover, they allow stronger unions in places like Europe to bring pressure to bear on TNCs for which their members work when these corporations seek to expand into parts of the world where labor organizations are not as strong. Wills (2002), for instance, has highlighted how European unionists convinced French hotel chain Accor not to oppose its employees' unionization efforts in countries such as Indonesia, New Zealand, and Ghana. There are myriad other examples of using such agreements for these purposes. What is significant about them is that they allow workers from one part of the globe to, in a sense, project geographically their strength to those places where unions have a harder time organizing. The agreements, in other words, are tools for spatially leveraging worker strength from one place to another.

While transnational entities like the GUFs and the ICTU have played important roles in linking together workers in different parts of the globe and in fighting for labor rights and better conditions, both for workers who belong to labor unions and for those who do not, individual unions and their national federations have also been involved in such activities. This international cooperation has a long history, especially in industries implicated in international transportation – presumably because the industries in which they labor are, by their very nature, focused upon the movement of goods through "international" space. For instance, some of the earliest international conferences of unionists took place in London in the 1890s, when seafarers and dockers from several nations came together to discuss issues of concern. US unions such as the International Ladies' Garment Workers' Union, the United Auto Workers (UAW), and the United Steelworkers of America (USWA) have also worked with unions in other countries since the earliest decades of the twentieth century. In the case of autoworkers, one of the most significant developments in their industry was the creation in the 1960s of a series of World Auto Councils, in which workers from Chrysler, Ford, and General Motors plants in fourteen countries joined together to share information about wage rates and working conditions, with the goal of facilitating a degree of international cooperation and, perhaps, ultimately coordinated bargaining. More recently, the UAW established a Council of Ford Workers covering Canada, the United States, and Mexico (Howard 1995). For their part, workers at the Mercedes-Benz plant in Alabama have worked with the German IG Metall union, which has pressured the company not to oppose unionization at its US facilities. IG Metall and other German and US unions also became heavily involved with unions in Eastern Europe as part of the economic and political transition there following the collapse of communism (Herod 2001), while the US Service Employees' International Union has sent organizers to coordinate with unions representing janitors and others in Australia, New Zealand, the United Kingdom, Poland, India, France, Switzerland, Germany, the Netherlands, and South America (Stern 2006: 112). Meanwhile, Britain's GMB union has worked with Poland's Solidarność union to organize Polish workers in Britain and has even established Polish-speaking branches in Southampton and Glasgow (*Guardian* 2006).

Such have been the connections between unions in different countries that several have explored the possibility of formally merging. In the early 2000s, the British Amalgamated Engineering and Electrical Union and Germany's IG Metall discussed a possible merger that would have created a pan-European union representing more than 3 million manufacturing workers. In 2008, the USWA and Unite, which represents workers in the manufacturing, transportation, and service sectors and is the largest trade union in the United Kingdom and Ireland, agreed to merge to form a trans-Atlantic union called Workers Uniting. With about 3 million members in the steel, paper, oil, healthcare, and transportation industries, Workers Uniting largely came about as a way to synchronize pay talks and address issues of growing precarity by holding trans-Atlantic negotiations with multinational companies like oil conglomerate BP and ArcelorMittal, the world's largest steel maker. As Leo Gerard, USWA president noted, such a merger is designed to give workers greater strength to shape how the global economy unfolds: "Globalisation," he said, "has given financiers licence to exploit workers in developing countries at the expense of our members in the developed world. Only global solidarity among workers can overcome this sort of global exploitation wherever it occurs." Derek Simpson, head of Unite, likewise recognized how TNCs can use their geographical organization and the fact that they are spread across multiple countries to their advantage and that merging transnationally can give workers a means to counter this. He declared: "The political and economic power of multinational companies is formidable. They are able to play one nation's workers off against another to maximise profits. They do the same with governments, hence the growing gap between the rich and the rest of us. With this agreement we can finally begin the process of closing that gap" (*Financial Times* 2008).

In considering how labor organizations challenge the power of TNCs and other firms to shape local and national labor markets and the global economy, it is also important, however, to recognize that unions frequently also work hand-in-hand with their employers as a way to secure jobs for their members, often at the expense of workers in other countries. Indeed, sometimes it is hard to determine unions' exact motives when they engage in international solidarity actions; they may suggest that they are acting internationally as a way to help foreign workers – and they often do so, even when it is against their own interests and/or brings them no material return (for examples, see Ahlquist and Levi 2013) – when, in actuality, their efforts can also sometimes be designed to improve these foreign workers' wages and working conditions to such a degree that they are no longer attractive to relocating capital and therefore not a threat to the union's members. Johns (1998) calls these kinds of activities "accommodationist solidarity" actions, because they essentially seek to accommodate the extant power relations of global capitalism rather than to challenge them – she calls actions of this latter type "transformatory solidarity." One such example among many in this regard is that of several US unions and the AFL, which have a long record of working with the US government and corporations to undercut more radical unions in various parts of the world. Their efforts were tied up in the politics of the Cold War and the belief that unions which opposed US interests must, by definition, be sympathetic to the Soviet Union. Given its geographical proximity, US unions were mostly engaged in Latin America and the Caribbean, although they have also helped expand US interests in other parts of the globe. The main belief driving such activities (in addition to a concern about the spread of communism) was the idea that helping US businesses expand overseas, even if that required undermining local unions, would be good for US workers – at a time when the United States was the manufacturing workshop of the twentieth-century world, what was good for US business overseas was often seen as being good for US workers, because expanded overseas markets would mean greater exports from the United States and therefore more jobs at home. As part of their effort to inculcate pro-American attitudes among workers in other parts of the world, US unions often funded the construction of schools, community centers, and local infrastructure projects like bridges and irrigation ditches, together with housing for members of unions deemed supportive of US interests. In the case of Latin American and the Caribbean, through its American Institute for Free Labor Development, the American Federation of Labor–Congress of Industrial Organizations (AFL–CIO) used its own and US government funds to build several thousand units of housing in the 1960s and 1970s in places like Mexico City, Buenos Aires, Bogotá, São Paulo, Ribeirão in Brazil's southern coastal sugar zone, San Pedro de Macorís in the Dominican Republic, Montevideo, and a host of other communities across the region (for more details, see

Herod 2001: ch. 6). In so doing, the US unions shaped how workers in these communities were able to reproduce themselves socially and biologically, which has had important consequences for their ability to thrive over the years.

Organizing in the Age of Precarity

Labor unions have successfully established structures through which they can exercise influence at a number of different geographical scales, from the very local to the truly global. Here, though, I want to explore some of the ways in which workers are addressing issues specifically of precarity, which is arguably one of the greatest challenges to their abilities to sustain themselves that they presently face. Some of these actions are undertaken individually – as when a textile worker who is unhappy with her conditions of work deliberately sews the wrong color button onto an item of clothing as a little act of rebellion (what Scott (1985) has called an instance of "small arms fire in the class war") – and some are taken collectively, either through collaboration with labor unions or through other avenues. Such actions are mostly taken at the local level, but they are sometimes much more geographically widespread and coordinated, such as the 2012 nationwide strike called by the Congress of South African Trade Unions to protest the contracting out of jobs through temporary employment firms (*New York Times* 2012). Occasionally, campaigns are even global in scope. Every October 7 since 2008, for instance, the ITUC, in collaboration with local organizations, has held a World Day for Decent Work. Suggesting that "No action is too big or too small[,] whether be it a round table discussion, a huge demonstration, a protest letter, a flash mob action or something completely different," the ITUC each year arranges events in places as diverse as Burkina

Faso, Zambia, Colombia, Brazil, Bangladesh, Japan, Bermuda, Moldova, Australia, Indonesia, and many others to press governments and employers to improve workers' conditions.

One of the key sectors in which organizations supporting workers' rights have been active has been the global textile and apparel industry. Indeed, this industry seems to have caught the global popular imagination in ways that many other industries have not. One reason for this could be because of a series of PR campaigns designed to get Global North consumers to think about the conditions suffered by the (usually) Global South workers who make the clothes on sale at places like Marks & Spencer, the GAP, Old Navy, C&A, Macy's, and other retail outlets. Or it could be a result of the global coverage of deadly accidents like the May 2013 collapse of the eight-story Rana Plaza commercial building in Dhaka, Bangladesh, which left an official death toll of 1,129 and a further 2,500 injured and which is thought to be the deadliest garment factory accident in history. Whatever the reason, labor unions in these sectors have been energetically engaged in trying to improve their members' conditions. In light of the Rana Plaza disaster, for instance, more than fifty manufacturers agreed to sign a legally binding building safety agreement backed by the IndustriALL GUF and the Bangladeshi government (*Guardian* 2013). Lots of other entities have also been involved. One, the Asia Floor Wage Alliance (AFWA), has been waging a Clean Clothes Campaign for several years under the banner "Stitching a decent wage across borders," with the goal of developing a regional collective bargaining strategy that, they hope, will counter the threat of capital mobility, will prevent competition based on wage levels between Asian garment exporters (i.e., prevent a so-called "race to the bottom," where workers simply undercut each other in terms of the wages for which they are willing to work), and will make sure that profits are shared along the supply chain. By focusing upon the large, first-tier manufacturers who have direct supply relations with major brands and retailers (which are themselves often TNCs), the AFWA hopes to force them to bring pressure to bear on second-, third-, and even fourth-tier subcontractors, who are often difficult for labor activists to locate (Merk 2009). Significantly, this tactic – of bringing pressure to bear on retailers rather than trying to organize garment factories directly, an almost impossible task in an industry in which capital and equipment (which is often little more than a sewing machine) are frequently extremely foot-loose – has also been followed by activists in the United States, as they have fought the spread of sweatshops (Johns and Vural 2000). The Institute for Global Labour and Human Rights (known until 2011 as the National Labor Committee), for example, has promoted better conditions for garment workers through using the power of the press, as US TV personality Kathie Lee Gifford found out to her chagrin in 1996 when it announced that clothing bearing her name was being made by children in sweatshops in Latin America. The subsequent furor led then-President Bill Clinton to announce a plan by which major produc-ers like L.L. Bean, Liz Claiborne, Phillips-Van Heusen, and others would begin labeling (though only voluntarily) their garments as free of child and forced labor. Eventually, the actions of the NLC and the Clinton Administration led to the creation of the Fair Labor Association, a nonprofit entity that monitors conditions along various commodity chains in the clothing and footwear industries. Given these sub-sequent developments, it is often argued that the Gifford campaign was the pivotal event in bringing labor abuses in the garment industry to the attention of the US mainstream media. One important group to emerge in this context has been the United Students Against Sweatshops, formed in 1998 and now with chapters at more than 250 universities and colleges in North America, which lobbies against labor abuses in the collegiate apparel market.

Although it often proves difficult to organize precarious workers because much of what they do, by definition, takes place in the shadows of the economy, some unions have nevertheless attempted to do so. For instance, in the United States, the Communications Workers of America and the United Steelworkers have reached out to call center workers, as have, elsewhere, the Australian Service Union and Germany's Vereinte Dienstleistungsgewerkschaft (ver.di). In India, the Union for Information Technology & Enabled Services Professionals (UNITES Pro) has been working to improve conditions for call center workers, both current and future (the latter through linking with youth organizations and schools) (for more on unionizing Indian call centers, see Taylor and Bain 2010). Likewise, in South Korea the unions have organized strikes and sit-ins against large firms (*chaebols*) like Samsung and Hyundai to force them to stop using contract workers, who often find themselves working alongside directly employed workers on the same assembly line (for examples from the auto industry, see Yun 2016). However, in many other instances, precarious workers have turned to other mechanisms by which to improve their working conditions. In 2000, Microsoft paid US$97 million to settle an eight-year-long class action lawsuit brought on behalf of 8,000–12,000 permatemps who claimed that they were really permanent employees and that the company had improperly denied them benefits to which they would have been entitled had they been directly hired (*New York Times* 2000). In 2005, FedEx drivers in California sued the company, claiming that under

California law they were direct employees rather than independent contractors, a claim upheld by the US Court of Appeals for the 9th Circuit (*Alexander et al. v. FedEx Ground Package System, Inc.*, Nos. 12-17458, 12-17509 [9th Cir. Aug. 27, 2014]). The Supreme Courts of Kansas and of Missouri have also recently issued rulings challenging companies' designations of their employees as "independent contractors," while in 2011 the US Department of Labor launched a "Misclassification Initiative" and signed a "Memorandum of Understanding" with fourteen states that was designed to narrow the grounds on which workers can be classified as independent contractors (Blaisdell 2014). In 2016, the ride service Uber agreed to pay 385,000 drivers in California and Massachusetts nearly US$100 million to settle a class action law suit, although a federal judge later rejected the settlement on the grounds that the figure was too low (*Los Angeles Times* 2016). In October 2016, in a decision that could affect thousands of British Uber drivers, an employment tribunal ruled that Uber had unfairly treated two of its drivers when it considered them self-employed rather than employees of the company (Hodges and Kahn 2016). Several other such "gig employers" have also been sued in the United States and elsewhere (Kessler 2015), while Uber's low-cost UberPop service has been banned in France and the Netherlands for using unlicensed drivers.

Workers and their supporters have engaged in other forms of action designed to improve their situations. One innovative strategy has been to lobby for the implementation of "living wage" laws. These typically require that firms that are recipients of public money (in the forms of contracts for services they provide) pay their workers a living wage, which in most cases is much higher than the prevailing minimum wage set by law. They follow from a long tradition of cities enacting wage laws – in the 1960s, for instance, Baltimore, New York City, and Washington, DC enacted minimum wage ordinances that provided for higher minimum wages than those required by state legislation. However, the fact that workers earning the minimum wage might still live in poverty, even though they worked full time, led some cities to consider living wage laws. The first US city to pass such a law was Des Moines, Iowa, in 1988, though Baltimore's 1994 law was the first to use the term "living wage" explicitly (Luce 2004). Since then, dozens more cities and counties have followed suit. Although these laws are structured differently in different cities, often they include two levels of wages – one that is slightly lower if an employer also provides health insurance and one that is higher if that is not the case. Similar living wage campaigns have been pursued in other countries, including the United Kingdom, Canada, Australia, New Zealand, and South Africa, although their specifics vary; whereas US living wage campaigns have tended to focus upon government contracts, in London the campaign started in 2001 aimed to establish a living wage that would be paid by all employers in the city (Wills and Linneker 2012). What is significant about these campaigns is that they are not so much focused upon traditional collective bargaining structures linking employers and workers as on changing the geographical terrain of pursuing improved conditions for workers by requiring that living wages be paid across particular spatially defined labor markets – i.e., those circumscribed by the physical limits of the city within which they have been enacted. This also removes the focus of the employer–employee relationship from that of labor and employment law to that of bringing political pressure to bear upon local elected officials to pass living wage laws, a shift that potentially allows for a broadening of the struggle for improved conditions beyond the realm of the individual workplace to the wider community.

Perhaps one of the most revolutionary developments challenging precarity, however, has involved workers occupying their workplaces and engaging in forms of self-management. What is significant about these activities is that whereas the examples mentioned above concern workers trying to secure a better deal for themselves from their employers, efforts to occupy workplaces and engage in self-management are about workers freeing themselves of any relationship with an employer – that is to say, of making themselves precisely *not* a resource for capital to use but of becoming independent of it. Certainly, the tactic of occupying workplaces is not new. During the *Biennio Rosso* ("Two Red Years") of 1919 and 1920 in Italy, radicalized workers occupied hundreds of factories in the country's industrial northwest and established factory councils to run them. In the United States in the 1930s, auto workers engaged in sit-down occupations of plants, actions that eventually led to their employers' recognition of the United Auto Workers. In the late 1960s and early 1970s, a wave of plant occupations took place in Belgium to resist workers being laid off, including at an Anglo-German metalworks, a cloth factory, and the Brussels department store Union économique. During the restless summer of 1968, almost 9 million French workers (both unionized and non-unionized) occupied their workplaces, including at Sud Aviation and Renault. At a watch factory in Besançon the workers took over the plant and operated it for two years as a worker-managed plant under the slogan "C'est possible: on fabrique, on vend, on se paie!" ("It's possible: we make them, we sell them, we pay ourselves!"). In Britain, between 1971 and 1974, more than 200 workplaces were taken over by workers (Sherry

2010). Other experiments in worker self-management were also developed. For instance, when threatened with job loss at Lucas Aerospace, a military contractor, workers occupied the company's largest site in Burnley, in the north of England. Employees across the firm's seventeen constituent factories then developed a plan to shift production away from weapons and toward making more of the "socially useful" goods (like medical devices) that it also produced (Wainwright and Elliott 1982).

More recently, there has been a wave of factory occupations in response to contemporary economic problems across the planet. Arguably, the most famous of these have been the factory occupations in Argentina, undertaken as part of the umbrella Movimiento Nacional de Empresas Recuperadas (MNER – "National Movement of Recovered Companies"). With the collapse of the Argentine economy in 2001, workers increasingly began to take over factories and use them to produce goods that are socially necessary but not necessarily profitable to produce. Before too long, some 180 factories employing more than 10,000 workers had been reopened as worker-run cooperatives (Lavaca Collective 2007; Ranis 2005). In response to austerity policies and the hardships caused by the recent global economic recession, workers in Spain have also taken over plants to prevent job losses. Likewise, workers in Zrenjanin, Serbia (who oppose privatization of the Shinvoz plant that produces trains and locomotives), France (who sought to end the continued closure of an ArcelorMittal plant in Florange), Greece (where workers occupied the Vio.Me. building materials factory in Thessaloniki in 2013), as well as in the United States, Britain, Ukraine, Egypt, Turkey, South Korea, China, and hundreds of other places, have also taken over their workplaces, with many viewing their ownership and operation of these factories as an important counterpoint to the logic of neoliberal capitalism (Faulk 2008). Many workers have also become involved in cooperatives as a way to have greater control over their working conditions. Some of these are very small operations, but others are really quite large. The employee-owned John Lewis Partnership, for example, which was founded in 1929 as an "experiment in industrial democracy" (Lewis 1948) and which operates dozens of department stores and supermarkets, as well as other retail-related activities (Cathcart 2009), was the third-largest private company by sales in the United Kingdom in 2016 and provides its nearly 90,000 employees (who receive a share of annual profits) some significant say in running the business. For its part, the Mondragón Cooperative Corporation in northern Spain has more than 70,000 employees working in more than 100 separate worker-owned cooperative businesses in the industrial, retail, finance, and knowledge sectors of the economy (Whyte and Whyte 1991; Ormaechea 2003). Significantly, there have been efforts to link traditional labor union organizing strategies with the model of democratic economic control exhibited by cooperatives. In October 2009, for instance, the United Steelworkers of America announced an agreement with Mondragón to develop unionized worker cooperatives in the US manufacturing sector. Meanwhile, in some countries major politicians have pushed for worker cooperatives as a way to challenge the types of "crony capitalism" that they believe have been responsible for the recent global economic crisis and the precarity faced by workers. In 2012, the UK deputy prime minister, Nick Clegg, called for the creation of what he called a "John Lewis Economy," in which the market would become "a market for the many, not a market for the few" (*Guardian* 2012). These developments are potentially important both for workers and for the economy generally, as evidence suggests that because cooperatives' primary need is not to generate massive profits for shareholders but, instead, to provide a living for their members, they tend to be more productive and provide more stable employment for workers (Pérotin 2016). It also suggests that worker/community-controlled production is often more environmentally conscious than is production purely in pursuit of maximum profitability, which has important consequences for ecological sustainability and, thus, in reducing workers' biological precarity (see Chapter 4).

A final strategy pursued by some labor groups with regard to challenging corporate decision-making concerning investment decisions has been what we might call a "follow the things" approach to understanding the unfolding geography of global capitalism. This allows labor activists and unions to see how they are linked to workers in other parts of the globe through their labor processes and what they produce, with the result that it can help them to develop trans-spatial solidarities and, thereby, to shape the geography of commodity chains and the movement of economic value through them. In North America in the 1990s, for instance, a group of critical scholars and activists established the International Research Network on Autowork in the Americas to map the links between various suppliers and manufacturers, a project they labeled the "Mapping Supplier Chains" project. Activists have attempted to trace other types of commodities, such as the minerals used in manufacturing cell phones and other electronics, as a way to bring pressure to bear upon companies like Apple, Microsoft, Hewlett-Packard, and others. Although this type of activity typically rests within a particular sector of the economy – say, manufacturing, in the case of following the movement of automobile components – at least one innovative approach has tried to follow a commodity across sectoral divisions, from its very start to its final consumption. Hence, the International Union of Food and Allied Workers' Associations GUF, supported by the Dutch Federatie Nederlandse Vakbeweging labor federation, has tried to organize along the cocoa commodity chain by linking West African plantation workers with workers in European chocolate factories and shop workers who sell the final product. The variety of working conditions along the chain have made it difficult to achieve many practical results, but the exercise has nevertheless proven useful for educating European confectionary workers about conditions in West Africa's cocoa plantations and has resulted in some efforts to combat the kinds of child slavery detailed in Chapter 5.

Summary

Rather than remaining passive in the labor market, workers have clearly sought to improve their conditions through their actions. Whether engaging in individual acts of resistance to their employers' demands, coming together in labor unions to challenge their employers' prerogatives in the workplace, bringing pressure to bear on local elected officials to enact living wage laws, using the courts to contest their status as contingent workers, or engaging in a multitude of other activities, working people have played important roles in shaping how labor markets operate in different places at different times. Some have even gone so far as to withdraw from the employer–employee relationship entirely, by establishing worker cooperatives and occupying abandoned factories. Certainly, working people have not been entirely free to act as they please in this regard, for their abilities to challenge their positions in the labor market have varied over time and across space. However, the fact that they have engaged in activities like those described above has important implications for both labor as a resource and for how the economic geography of capitalism – and other systems of organizing work – is made. Through their agency, then, workers transform both themselves and their surroundings. In so doing, they reveal themselves to be unique amongst resources.

CHAPTER EIGHT

Concluding Thoughts

In this book I have explored a number of issues that are important to consider when reflecting upon labor as a resource. Chapter 1 detailed some of the reasons why labor is quite different from other resources and why its geographical location is especially important to its nature and behavior. Chapter 2 surveyed the historical geography of labor's distribution across the globe, a distribution that is largely of workers' own making, both in terms of their past and present-day migration patterns and of the demographic processes at play in various parts of the planet at different times – processes over which they exert much (but not total) control. In Chapters 3 and 4, I looked at some of the impacts that two powerful contemporary processes – globalization and neoliberalization – are having on workers' lives. Significantly, both processes have specific histories and geographies to them, affecting various workers and peasants in different ways, depending upon where they are located and how they are embedded in individual places and are linked across space (or not). Equally significantly, though, through their actions, working people are shaping how both globalization and neoliberalization themselves are unfolding; engaging in transnational solidarity campaigns and challenging efforts to undermine work security are just two of the ways in which they are doing so. As we have also seen, what is important about the impact of globalization and neoliberalization on economies and the people who toil in them is that some of the changes in the political economy of contemporary capitalism that have been linked to globalization and neoliberalization – like the ever greater use of automation and computerization and the growth of flexible work arrangements – have led many commentators to suggest that we are transitioning from an Old Economy of labor-intensive manufacturing to a New Economy of knowledge work, with obvious implications for working people. However, rather than seeing a significant disjuncture between, on the one hand, an Old Economy that is redolent of the nineteenth and twentieth centuries, one built upon routinized mass production and sweaty and dirty drudge work, and, on the other, a New Economy of salubrious knowledge work that will emancipate workers as we move further into the twenty-first century, what we are actually seeing are substantial historical and geographical continuities between the past, the present, and what is likely to be the future – as Chapters 5 and 6 both illustrated, Old Economy work practices are widespread in both Old *and* New Economy sectors today. Finally, in Chapter 7 we saw some of the ways in which working people have proactively transformed the conditions under which they labor, through creating institutions like labor unions and various other means. These efforts, too, have been shaped by how workers are embedded in particular places and how they are connected (or not) over space – that is to say, there is a historical geography to these efforts and their success or failure.

The book, then, has detailed some of the diverse working experiences across the planet which millions of workers encounter. Throughout, the key point that I have articulated is that labor is a resource unlike any other because of its sentience and ability to proactively shape both the conditions under which it is used by its employers and the spatial contexts within which it is used. However, while I have primarily talked about labor as a resource in the context of the production of goods and services, it is also important to be mindful of the fact that working people live their lives not just as producers, but also as consumers of goods and services. This, too, makes them a resource unlike others, as it is through their conscious consumption choices that they facilitate manufacturers' and service providers' realization of profit – making a profit requires not just the production of goods or services, but also their sale to others. As with the realm of production so with that of consumption, though: working people's consumption patterns have particular historical geographies over which they have a fair amount of, but not total, control – what goods and services they consume, and where and how they do so, are shaped to a great degree by their own desires, as well as by other actors and forces (employers' low wages can put limits on their disposable income, poor weather can restrict the amount of food they have to consume, manufacturers may stop producing certain goods which some people had long consumed, and so forth).

As we have seen in many of the examples discussed in this book, workers can proactively shape when, where, and how their labor is employed and when, where, and how it consumes other resources and goods and services. They can even resist their use by others, by refusing to work and/or consume particular things – the potential power of labor in this latter regard is highlighted by the fact that consumer boycotts are often a popular action undertaken against producers who are deemed not to be treating their employees or others sufficiently well. However, when contemplating how labor as a resource is both an object utilized by others and also its own subject, capable of transforming the conditions under which it is used, as well as how it is a resource that is both trapped by its own geography and also capable of proactively altering that geography to better serve its own interests, it is imperative to remember that although workers have some degree of agency in shaping how they are used and can behave, they do not have complete free rein. As Marx (1852 [1963]) long ago outlined, their ability to make their own histories – and, I would add, geographies – is constrained by the very histories and geographies which have shaped their existence.

One way to conceptualize the degree of power which workers possess to shape the economic and socio-spatial contexts within which they find themselves is through sociologist Erik Olin Wright's (2000: 962) distinction between "associational power" and "structural power." For Wright, structural power is the "power that results simply from the location of workers within the economic system." Silver (2003: 13) breaks this category down further, into what she calls "marketplace bargaining power," which is the power that "results directly from tight labor markets," and "workplace bargaining power," which is the power that results "from the strategic location of a particular group of workers within a key industrial sector." Such structural power contrasts with associational power, which results "from the formation of collective organizations of workers," including labor unions, political parties, works councils, cooperatives, and/or community organizations. What this distinction allows us to do is think about how the proactive actions of workers (such as coming together as part of a union or through occupying a factory) and the circumstantial aspects of workers' lives (the fact that they may be fortunate enough to find themselves in demand by employers in a labor market that is tight at particular times in particular places, but may not be so at other times/places) can both shape, though in different ways, labor's use as a resource. However, it also provides a way of escaping the simple assumption that well-organized workers (say, those in a union) automatically have greater capacity to shape how they are used as a resource than do those who are more precariously employed. One very brief example that highlights this point comes from recent worker struggles in Greece, which has been suffering significant economic hardships since the emergence of the government debt crisis in 2009 (for more on this crisis, see Mavroudeas 2013). Thus, whereas steelworkers who were members of the Greek metalworkers' union were unable to prevent the closure of several mills and the loss of hundreds of jobs through a series of protests and strikes in 2011–12, one group of extremely precarious workers – immigrant, non-union strawberry pickers – who might normally have been assumed to have had little power to change the conditions under which they were employed (i.e., to act as the subjects of their own history and geography) were, in fact, by abandoning the fields in which they labored, able to win a significant victory against farmers who had refused to pay them their wages. Without these immigrant workers' labor, the farmers who employed them would have lost their very valuable crops, which would have been left to rot in the fields, and they thus had little choice but to pay the strawberry pickers their withheld wages, an act that marked a remarkable success for the workers (for more details on these two disputes, see Gialis and Herod 2014). What this example illustrates is the significance of context in thinking about labor's agency and ability to shape the conditions under which it is used as a resource by others. Specifically, what gave the tremendously precarious immigrant strawberry pickers their power was not that they had developed formidable institutions but the fact that, at the very moment they were most needed to gather in the harvest, they walked off the job. At any other time of the year, they would have been relatively powerless. At harvest time, however, they could threaten their employers' economic futures by withholding their labor, an act that gave them great leverage.

In contemplating labor's role in shaping the conditions under which it is used as a resource, it is useful, finally, to consider Katz's (2004) disaggregation of labor agency into what she calls acts of "resistance," of "reworking," and of "resilience." For Katz, resistance is a form of agency that reflects an advanced consciousness concerning employers' oppressive power, one that results in workers seeking to challenge such power. Reworking, by way of contrast, is an intermediate form of agency, one that does not directly contest employers' power in the labor market, but that does seek to create a different balance of power, one more favorable to workers' interests. Finally, resilience is a more individualistic form of agency, one in which workers simply accept their position and do not really challenge the existing social order. Through contemplating these questions – different types of agency, the significance of contextual factors like the timing of disputes, whether workers have associational and/or structural power, how workers' geographical embeddedness affects their abilities to challenge the conditions under which they are hired, and so forth – we can gain a much more nuanced understanding of labor as a resource, a resource that is both used by others but is also capable of shaping profoundly the conditions under which this use occurs.

Notes

1 A RESOURCE UNLIKE ANY OTHER

1 More precisely, it is workers' capacity to labor that can be bought
  and sold, unless we are talking about slaves, whose bodies can
  indeed actually be purchased.
2 In his tract *The Eighteenth Brumaire of Louis Bonaparte*, Marx
  (1852[1963]: 15) famously stated: "Men make their own history,
  but they do not make it as they please; they do not make it under
  self-selected circumstances, but under circumstances existing
  already, given and transmitted from the past."

2 LABOR IN GLOBAL CONTEXT

1 Some of the decrease in urban population may reflect reforms
  that came after the 1789 revolution – the freeing of rural
  peasants from many taxes and duties to which they were
  historically obligated may have encouraged some return
  migration from cities but also likely reduced rural to urban
  migration, such that, as France's overall population grew (from
  about 21 million in 1700 to nearly 30 million in 1800), the
  proportion living in urban areas was reduced.
2 The GCC member states are Bahrain, Kuwait, Oman, Qatar,
  Saudi Arabia, and the United Arab Emirates (UAE).
3 In 1980, China's median age was 21.7 years of age; by 2030 and
  2050, it is expected to be 43.2 and 49.6 respectively (United
  Nations 2015b: 33).

3 GLOBALIZATION AND LABOR

1 £173 million from 1855 would be worth about £450 billion in 2016; £173 million from 1914 would be worth about £132 billion (Officer and Williamson 2016).

2 Unless otherwise indicated, the data in this paragraph come from UNCTAD (2016b).

3 JIT methods of production and inventory control involve delivering components to a factory just moments before they are used. This provides significant savings to the firms, including not having to build large warehouses in which to store components, having faster capital turnover times, and being able to detect defects in components sooner than if they were stored for months in a warehouse.

4 NEOLIBERALISM AND WORKING PRECARIOUSLY

1 "Deregulating" is in quotation marks to signal that it is a problematic term. Typically, people who use the term – especially those who advocate the "free market" – do so to suggest that there are fewer regulations guiding an economy after it has been "deregulated" than there were before. In reality, there are often actually many more regulations, as "deregulation" requires new sets of government rules – the "deregulation" of financial markets, for instance, does not mean that they will now operate without rules but, rather, that financial actors must operate according to different rules (for more on this point, see Vogel 1996).

2 Named for US engineer Frederick Winslow Taylor, Taylorism refers to the principles of "scientific management" that he developed through analyzing workflows within manufacturing industries in the 1880s and 1890s. Through better managing workflow, Taylor sought to improve labor productivity.

3 In the United States, although some 3.5 million manufacturing jobs will likely need to be filled during the next decade, it is estimated that about 2 million of these will remain vacant due to a lack of skilled workers (Deloitte 2015).

5 FROM DRUDGE WORK TO EMANCIPATED WORKERS?

1 US air force drones patrolling the skies of Afghanistan are actually piloted from an airbase just outside Las Vegas.

6 MEET THE NEW ECONOMY – SAME AS THE OLD ECONOMY?

1 In many ways, this utopian idea is quite old. In his 1930 essay "Economic Possibilities for our Grandchildren," British economist John Maynard Keynes famously predicted that by 2030 advanced industrial societies would be sufficiently wealthy that people would spend more time engaged in leisure pursuits than they would working, toiling for perhaps only 15 hours a week.
2 The term *bracero* means "one who works using his arms," and has come to refer to any kind of day laborer.
3 A list of call center locations around the world can be found at www.elsnet.org/ccorglist_c.html.
4 In an attempt to address this, in 2014 the Filipino construction firm Megaworld announced that it would spend some US$5 billion to construct ten mini-cities for BPO operations, with such mini-cities operating on the US sleep/wake cycle (Winn 2014).

Selected Readings

The topic of labor as a resource is potentially a huge one. Consequently, any effort to point you in the direction of additional selected readings is bound to be partial. So, in what follows, I will try to highlight what I think are some helpful readings, but you should recognize that these are not the only ones that you may find helpful.

Chapter 1 focuses upon how labor is a resource different from others. One of the first in-depth treatises on labor was Adam Smith's 1776 *An Inquiry into the Nature and Causes of the Wealth of Nations*, in which he argued that labor was the source of all wealth. Paul McNulty's overview of his views was written more than four decades ago, but it is nevertheless still useful: "Adam Smith's concept of labor," *Journal of the History of Ideas* 34.3 (1973): 345–366. In *On the Principles of Political Economy and Taxation* (1817), David Ricardo argued that labor was the origin of all wealth and he also developed the notion of comparative advantage. Karl Marx's three-volume epic *Capital* (1867) detailed the role played by labor under capitalism; it is a hefty read. A helpful guide to understanding these volumes is David Harvey's *A Companion to Marx's Capital* (London: Verso, 2010), while his *The Limits to Capital* (Oxford: Blackwell, 1982) attempts to spatialize Marx's thinking. For an overview of the field of labor economics, see *Labor Economics* by Pierre Cahuc, Stéphane Carcillo, and André Zylberberg (2nd edn., Cambridge, MA: MIT Press, 2014). Doreen Massey's

Spatial Divisions of Labor (New York: Methuen, 1982) is also an important book that seeks to spatialize understandings of labor, as is *Spaces of Work* by Noel Castree, Neil Coe, Kevin Ward, and Michael Samers (London: Sage, 2004). My own *Labor Geographies* (New York: Guilford, 2001) makes links between labor's agency and its spatiality, which has implications for thinking about labor as a resource.

Chapter 2 explains labor's current distribution across the planet. Part of this results from the migration of people from one place to another and part from *in situ* demographic processes. Empirical analyses of the history and geography of human migration are legion, but there are several texts that provide good overviews of how to conceptualize such movement. In *Labor Movement* (Oxford: Oxford University Press, 2006), Harald Bauder shows how migration patterns and the kinds of jobs that immigrants end up doing in their destination countries are deeply shaped by cultural, legal, and institutional processes. As his book's subtitle suggests, Khalid Koser's *International Migration: A Very Short Introduction* (2nd edn., Oxford: Oxford University Press, 2016) provides a brief overview to questions of international migration. Sam Scott's "Labour, migration and the spatial fix: Evidence from the UK food industry," *Antipode* 45.5 (2013): 1090–1109, outlines a conceptual framework for understanding migrant workers' impacts upon local economic development, while Michael Samers's "'Globalization', the geopolitical economy of migration and the 'spatial vent'," *Review of International Political Economy* 6.2 (1999): 166–199, explores how countries solve issues of domestic economic crisis through importing or exporting migrants. For an overview of demographic issues facing different parts of the world and how to conceptualize them, see John Weeks, *Population: An Introduction to Concepts and Issues* (12th edn., Boston: Cengage Learning,

2016). For a nontechnical assessment of future demographic change and its likely impacts, see Sarah Harper's *How Population Change Will Transform Our World* (Oxford: Oxford University Press, 2016).

The issue of globalization addressed in Chapter 3 is one upon which many writers have expressed their opinions. Amongst those most associated with the hyperglobalist school of thought is Japanese management guru Kenichi Ohmae, whose books *The Borderless World: Power and Strategy in the Interlinked Economy* (New York: HarperBusiness, 1990) and *The End of the Nation State* (New York: McKinsey and Company, 1995) argue that we are rushing headlong into an ever more globalized world. A counter-argument is made by Paul Hirst and Graham Thompson in *Globalization in Question* (2nd edn., Cambridge: Polity, 1999), who maintain that many of the proclamations made by people like Ohmae are overblown. The classic treatise on global production networks is that by Gary Gereffi and Miguel Korzeniewicz in *Commodity Chains and Global Capitalism* (Westport, CT: Praeger, 1994). One of the first articles to address the phenomenon of global destruction networks is Andrew Herod, Graham Pickren, Al Rainnie, and Susan McGrath-Champ, "Global destruction networks, labour, and waste," *Journal of Economic Geography* 14.2 (2014): 421–441. *The SAGE Handbook of Globalization* (2 vols., London: Sage, 2014), edited by Manfred Steger, Paul Battersby, and Joseph Siracusa, and *The Handbook of Globalisation, Second Edition* (Cheltenham: Edward Elgar, 2011), edited by Jonathan Michie, both provide a wealth of essays concerning globalization and its dynamics. The World Bank (http://data.worldbank.org) and United Nations Conference on Trade and Development (http://unctad.org/en/pages/home.aspx) have a plethora of statistics on foreign direct investment.

With regard to Chapter 4, Guy Standing's *The Precariat: The New Dangerous Class* (London: Bloomsbury Academic, 2011) is a must read. Rob Lambert and Andrew Herod's edited *Neoliberal Capitalism and Precarious Work* (Cheltenham: Edward Elgar, 2016) provides insights into how precarity is experienced by working people from several countries. Wayne Lewchuk, Marlea Clarke, and Alice de Wolff's *Working Without Commitments* (Montreal: McGill-Queen's University Press, 2011) examines precarious work's impacts upon workers' health. *Globalization and Precarious Forms of Production and Employment*, edited by Carole Thornley, Steve Jefferys, and Beatrice Appay (Cheltenham: Edward Elgar, 2010), links precarity with globalization and details some of the challenges it poses to labor and its organizations. Arne Kalleberg's *Good Jobs, Bad Jobs* (New York: Russell Sage, 2011) looks at the rise of precarious employment in the United States. An insightful book on low-wage work based on the author's first-hand experience is Barbara Ehrenreich's *Nickel and Dimed* (New York: Henry Holt, 2001). David Harvey's *A Brief History of Neoliberalism* (Oxford: Oxford University Press, 2005) and Manfred Steger and Ravi Roy's *Neoliberalism: A Very Short Introduction* (Oxford: Oxford University Press, 2010) provide very readable overviews of neoliberalism and how it has impacted different parts of the globe. The *Handbook of Neoliberalism* (Oxford: Routledge, 2016) edited by Simon Springer, Kean Birch, and Julie MacLeavy, on the other hand, is a large tome that explores several aspects of neoliberalism. Henk Overbeek and Bastiaan van Apeldoorn's *Neoliberalism in Crisis* (Basingstoke: Palgrave Macmillan, 2012) surveys challenges that neoliberal capitalism has faced in the years since the 1997 Asian financial crisis.

Chapters 5 and 6 focus upon the continuities between the so-called "Old Economy" and "New Economy." Alan

Webber's "What's so new about the new economy?," *Harvard Business Review* 71.1 (1993): 24–42, provides a readable early overview of the supposed differences between the Old and New Economies. Erik Brynjolfsson and Andrew McAfee's *Race Against the Machine* (Lexington: Digital Frontier Press, 2011) looks at technology's impact upon work and labor markets. For further reading on the economic sectors mentioned in these two chapters, probably the finest work on labor relations in Western Australia's Pilbara is Bradon Ellem's *Hard Ground* (Port Hedland: Pilbara Mineworkers Union, 2004). His "Geographies of the labour process: Automation and the spatiality of mining," *Work, Employment and Society* 30.6 (2016): 932–948, brings much of the story up to date. Hugh Hindman's *The World of Child Labor* (London: M.E. Sharpe, 2009) and Kevin Bales's *Disposable People* (London: University of California Press, 2012) both detail some of the slave-like conditions faced by many workers – especially children – in the contemporary global economy. The Asia Foundation/International Labour Organization's 2015 report *Migrant and Child Labor in Thailand's Shrimp and Other Seafood Supply Chains* provides a comprehensive overview of labor conditions in the Southeast Asian fishing industry. *Challenging the Chip* (Philadelphia: Temple University Press, 2006), edited by Ted Smith, David Sonnenfeld, and David Pellow, contains several chapters detailing hi-tech workers' plight, while *The Silicon Valley of Dreams*, by David Pellow and Lisa Park (New York: NYU Press, 2002), explores issues of immigration and environmental racism in Silicon Valley. A special issue of *European Journal of Work and Organizational Psychology* 12.4 (2003): 305–427 contains articles detailing conditions faced by call center workers, as does Emily Yellin's *Your Call Is (Not That) Important to Us* (New York: Free Press, 2009). Phil

Taylor and Peter Bain's "Call centre offshoring to India: The revenge of history?," *Labour & Industry* 14 (2004): 15–38, was one of the first articles to explore the phenomenon of the Indian call center. For more on the life of nannies and cleaners, see *Global Woman*, edited by Barbara Ehrenreich and Arlie Russell Hochschild (New York: Metropolitan Books, 2003), and *The Dirty Work of Neoliberalism*, edited by Luis Aguiar and Andrew Herod (Oxford: Blackwell, 2006).

Finally, Chapter 7 details workers' efforts to self-organize. A good overview of how workers have sought to come to terms with globalization during the past 150 years is Beverly Silver's *Forces of Labor* (Cambridge: Cambridge University Press, 2003). The edited collection *Global Labour History* by Jan Lucassen (New York: Peter Lang, 2008) contains essays that detail how to go about studying labor and provide an overview of labor's experience in several parts of the globe. John Windmuller, Stephen Pursey, and Jim Baker provide a good institutional history of worker internationalism in "The international trade union movement," in *Comparative Labour Law and Industrial Relations in Industrialized Market Economies*, edited by Roger Blanpain (The Hague: Kluwer Law International, 2004). An important early piece on the role that the state plays in shaping working relations is Gordon Clark's "A question of integrity: The National Labor Relations Board, collective bargaining and the relocation of work," *Political Geography Quarterly* 7.3 (1988): 209–227. Philip Cooke's "Class practices as regional markers: A contribution to labour geography," in *Social Relations and Spatial Structures*, edited by Derek Gregory and John Urry (New York: St. Martin's, 1985), is an early contribution to what has become called labor geography, while Ray Hudson and David Sadler's "Contesting work closures in Western Europe's old industrial regions: Defending place or betraying class?" in *Production, Work, Territory,* edited by Alan Scott and Michael Storper (Boston: Allen & Unwin, 1986), explores the tensions between workers' class and geographical interests. Don Mitchell's *The Lie of the Land* (Minneapolis: University of Minnesota Press, 1996) is an important book that explores migrant workers' roles in making the California landscape, while the collections edited by Andrew Herod, *Organizing the Landscape* (Minneapolis: University of Minnesota Press, 1998), and Susan McGrath-Champ, Andrew Herod, and Al Rainnie, *Handbook of Employment and Society* (Cheltenham: Edward Elgar, 2010), contain a multitude of chapters detailing workers' efforts to organize in different contexts. Finally, my own bibliography "Geography of Labor" for Oxford Bibliographies, available online at http://tinyurl.com/jgp-sjby, has several hundred annotated readings relating to labor organizing.

References

ABS (2013) *Towns of the Mining Boom*. Australian Bureau of Statistics: Canberra, Australia.

ABS (2015) *Characteristics of Employment, Australia, August 2014*. Australian Bureau of Statistics: Canberra, Australia.

Afoakwa, E.O. (2016) *Chocolate Science and Technology*, 2nd edn. Wiley-Blackwell: Chichester, UK.

Agger, B. (1989) *Fast Capitalism: A Critical Theory of Significance*. University of Illinois Press: Urbana, IL.

Aguiar, L.L.M. (2001) "Doing cleaning work 'scientifically': The reorganization of work in the contract building cleaning industry." *Economic and Industrial Democracy* 22.2: 239–269.

Aguiar, L.L.M. (2006) "Janitors and sweatshop citizenship in Canada." *Antipode* 38.3: 440–461.

Ahlquist, J.S., and Levi, M. (2013) *In the Interest of Others: Organizations and Social Activism*. Princeton University Press: Princeton, NJ.

Aliber, R.Z., and Click, R.W. (1993) *Readings in International Business: A Decision Approach*. MIT Press: Cambridge, MA.

Anderson, B. (2003) "Just another job? The commodification of domestic labor." In Ehrenreich, B., and Hochschild, A.R. (eds.), *Global Woman: Nannies, Maids, and Sex Workers in the New Economy*. Metropolitan Books: New York, pp. 104–114.

Apple Inc. (2016) *Environmental Responsibility Report: 2016 Progress Report, Covering Fiscal Year 2015*. Apple Inc.: Cupertino, CA.

Aronowitz, S. (1990) "Writing labor's history." *Social Text* 25/26: 171–195.

Arun, M.G. (2013) "Last call: Expensive manpower, inadequate language skills and poor infrastructure are forcing BPO businesses to leave India. Young job seekers are the victims." *India Today*, March 15. Available at http://indiatoday.intoday.in/story/bpo-industry-call-centre-culture-dying-in-india/1/258032.html.

Associated Press (2015) "Global supermarkets selling shrimp peeled by slaves." December 14. Available at http://bigstory.ap.org/articl e/8f64fb25931242a985bc30e3f5a9a0b2/ap-global-supermarkets-selling-shrimp-peeled-slaves.

Australian Government (2015) "The Australian gold rush." Available at www.australia.gov.au/about-australia/australian-story/austn-go ld-rush.

Australian Mining (2013) "Fortescue install automated trucks at its Solomon Hub Kings mine." October 4.

Bacon, D. (2015) "The maquiladora workers of Juárez find their voice." *The Nation*, November 20.

Baeninger, R., and Guimarães Peres, R. (2011) "Refugiados africanos em São Paulo, Brasil: Espaços da migração." *RILP: Revista Internacional em Língua Portuguesa* 24: 97–110.

Baily, S. (1969) "The Italians and the development of organized labor in Argentina, Brazil, and the United States: 1880–1914." *Journal of Social History* 3.2: 123–134.

Bairoch, P. (1982) "International industrialization levels from 1750 to 1980." *Journal of European Economic History* 11.2: 269–333.

Baldwin-Edwards, M. (2011) *Labour Immigration and Labour Markets in the GCC Countries: National Patterns and Trends.* Kuwait Programme on Development, Governance and Globalisation in the Gulf States, London School of Economics: London.

BBC News (2004) "Call of India lures European workers." November 24. Available at http://news.bbc.co.uk/2/hi/south_asia/4038069. stm.

Benner, C. (2002) *Work in the New Economy: Flexible Labor Markets in Silicon Valley.* Wiley-Blackwell: Chichester, UK.

Benner, C. (2007) "Regions and firms in eWork relocation dynamics: Pittsburgh's call centre industry." *World Organisation, Labour & Globalisation* 1.2: 98–115.

Berger, S. (2003) *Notre première mondialisation – Leçons d'un échec oublié.* Seuil: Paris.

Bidwai, P. (2003) "Call centres 'bad for India'." *BBC News*, December 11. Available at http://news.bbc.co.uk/2/hi/south_asia/3292619. stm.

Blair, T. (1998) *The Third Way: New Politics for the New Century.* Fabian Society: London.

Blaisdell, A. (2014) "Companies should take heed of recent cases finding workers improperly classified as independent contractors."

Available at www.greensfelder.com/employment-and-labor-blog/companies-should-take-heed-of-recent-cases-finding-workers-impro perly-classified-as-independent-contractors.

Bornschier, V. (2000) "Western Europe's move toward political union." In Bornschier, V. (ed.), *State-building in Europe: The Revitalization of Western European Integration.* Cambridge University Press: Cambridge, pp. 3–37.

Brayer, H.O. (1949) "The influence of British capital on the Western range-cattle industry." *Journal of Economic History* 9 (Supplement: *The Tasks of Economic History*): 85–98.

Browning, J., and Reiss, S. (1998) "Encyclopedia of the new economy: Redefining business for the 21st century." *Wired* 6.3: 105–114.

Burnham, L., and Theodore, N. (2012) *Home Economics: The Invisible and Unregulated World of Domestic Work.* Report for National Domestic Workers' Alliance: New York.

Butrica, B.A., Iams, H.M., Smith, K.E., and Toder, E.J. (2009) "The disappearing defined benefit pension and its potential impact on the retirement incomes of baby boomers." *Social Security Bulletin* 69.3: 1–27.

Carnoy, M., Castells, M., and Benner, C. (1997) "Labour markets and employment practices in the age of flexibility: A case study of Silicon Valley." *International Labour Review* 136.1: 27–48.

Cathcart, A. (2009) "Directing democracy: The case of the John Lewis Partnership." Unpublished PhD dissertation, School of Management, University of Leicester.

Chan, J. (2009) "Progress or illusory reform? Analyzing China's labor contract law." *New Labor Forum* 18.2: 43–51.

Chan, K.W. (2013) "China: Internal migration." In Ness, I. (ed.), *The Encyclopedia of Global Human Migration*, vol. 2. Blackwell: Malden, MA, pp. 980–995.

Charles, B., Rastogi, S., Sam, A.E., Williams, J.D., and Kandasamy, A. (2013) "Sexual behavior among unmarried business process outsourcing employees in Chennai: Gender differences and correlates associated with it." *Indian Journal of Public Health* 57.2: 84–91.

China Labor Watch (2012) *An Investigation of Eight Samsung Factories in China: Is Samsung Infringing Upon Apple's Patent to Bully Workers?* Report, dated September 4.

Choy, C.C. (2003) *Empire of Care: Nursing and Migration in Filipino American History.* Duke University Press: Durham, NC.

CIETT (2015) *Economic Report, 2015 Edition*. Confédération interna-
tionale des agences d'emploi privées/International Confederation
of Private Employment Agencies: Brussels.

Commission on Filipinos Overseas (2013) "Stock estimate of Filipinos
overseas as of December 2013." Available at www.cfo.gov.ph/down
loads/statistics/stock-estimates.html.

Constable, N. (2007) *Maid to Order in Hong Kong: Stories of Migrant
Workers*, 2nd edn. Cornell University Press: Ithaca, NY.

Corder, G.D., and Golev, A. (2014) "Industrial ecology forum 'Shifting
the Australian resources paradigm,' March 28, 2014, Sydney –
Outcomes and Findings Report." Prepared for Wealth from
Waste Cluster by the Centre for Social Responsibility in Mining,
Sustainable Minerals Institute, The University of Queensland,
Brisbane, Australia. Available at http://wealthfromwaste.net/wp-
content/uploads/2014/11/Industrial-Ecology-Forum-Shifting-The-
Australan-Resources-Paradigm_2014.pdf.

Corno, L., and de Walque, D. (2012) "Mines, migration and HIV/
AIDS in Southern Africa." *Journal of African Economies* 21.3:
465–498.

Cox, R. (2006) *The Servant Problem: Paid Domestic Work in a Global
Economy*. I.B. Tauris: London.

Crush, J. (1986) "Swazi migrant workers and the Witwatersrand gold
mines 1886–1920." *Journal of Historical Geography* 12.1: 27–40.

Daily Beast (2015) "Lawsuit: Your candy bar was made by child slaves."
September 30.

Daily Sabah (Istanbul) (2016) "Gulf countries employ highest number
of foreign workers." January 27.

Davies, J. (2005) *Hospital Contract Cleaning and Infection Control*.
Report by the Global Political Economy Research Group, School of
Social Sciences, Cardiff University, Wales.

Davis, K. (1965) "The urbanization of the human population."
Scientific American 213.3: 3–15.

Deloitte (2015) *The Skills Gap in US Manufacturing: 2015 and Beyond*.
Deloitte Development LLC: New York.

Department of Homeland Security (2014) *Yearbook of Immigration
Statistics: 2013*. US Department of Homeland Security, Office of
Immigration Statistics: Washington, DC.

Devreese, D.E. (1988) "An inquiry into the causes and nature of
organization: Some observations on the International Working
Men's Association, 1864–1872/1876." In van Holthoon, F. and

van der Linden, M. (eds.), *Internationalism in the Labour Movement 1830–1940, Volume 1*. E.J. Brill: London, pp. 283–303.

Drimie, S. (2002) "The impact of HIV/AIDS on rural households and land issues in Southern and Eastern Africa." A Background Paper prepared for the Food and Agricultural Organization, Sub-Regional Office for Southern and Eastern Africa. Available at www.fao.org/wairdocs/ad696e/ad696e00.htm.

ECCCO (2012) *European Contact Center Benchmark Platform 2012*. Report for the European Confederation of Contact Centre Organisations, Ottenburg, Belgium.

Eliott, M.A. (1916) *The Frozen Meat Industry of New Zealand*. H.L. Young: Palmerston North, NZ.

Ellem, B. (2015) "Resource peripheries and neo-liberalism: The Pilbara and the remaking of industrial relations in Australia." *Australian Geographer* 46.3: 323–337.

Environmental Justice Foundation (2013) *The Hidden Cost: Human Rights Abuses in Thailand's Shrimp Industry*. Environmental Justice Foundation: London.

EU-OSHA (2008) "Cleaners and musculoskeletal disorders." E-Facts 39, European Agency for Safety and Health at Work: Bilbao, Spain.

EU-OSHA (2009) "The occupational safety and health of cleaning workers." European Agency for Safety and Health at Work: Bilbao, Spain.

European Commission (2015) "Immigration in the EU." Available at http://ec.europa.eu/dgs/home-affairs/e-library/docs/infographics/immigration/migration-in-eu-infographic_en.pdf.

Eurostat (2015) "Part-time employment rate." Available at http://ec.europa.eu/eurostat/tgm/table.do?tab=table&init=1&plugin=1&pcode=tesem100&language=en.

Eurostat (2017) "Migration and migrant population statistics." Available at http://ec.europa.eu/eurostat/statistics-explained/index.php/Migration_and_migrant_population_statistics.

Fair Work Ombudsman (2011) *National Cleaning Services Campaign 2010–11: Report June 2011*. Australian Government: Canberra.

Fair Work Ombudsman (2016) "Cleaning industry compliance needs to improve." Press release, May 13. Australian Government: Canberra.

FAOStat (2016) Online Statistics Database, Food and Agriculture Organization of the United Nations Statistics Division. Available at www.fao.org/faostat/en/#data.

Faulk, K.A. (2008) "If they touch one of us, they touch all of us: Cooperativism as a counterlogic to neoliberal capitalism." *Anthropological Quarterly* 81.3: 579–614.

Feis, H. (1930) *Europe: The World's Banker, 1870–1914*. Yale University Press: New Haven, CT.

Financial Times (2008) "Unite becomes a transatlantic force." July 1.

Financial Times (2015) "The new world of work: Recovery driven by rise in temp jobs – non-permanent contracts can be a trap not a stepping stone." August 4.

Financial Times (2016) "Rio Tinto's driverless trains are running late: Delay to rail system in Australia will cut iron ore production." April 19.

Fortune (1996) "A call to profit: The technological revolution in call centers has stunningly transformed the way corporations large and small do business." Special advertising section. May 27.

Freedman, A. (1988) Testimony of Audrey Freedman before the Employment and Housing Subcommittee of the Committee on Government Operations, House of Representatives, 100th Congress of the United States, May 19.

Freelancers Union (n.d.) *Freelancing in America: A National Survey of the New Workforce*. Freelancers Union and Elance-oDesk: New York.

Frey, C.B., and Osborne, M.A. (2013) *The Future of Employment: How Susceptible are Jobs to Computerisation?* Oxford Martin Programme on Technology and Employment, University of Oxford: Oxford.

Gallup (2014a) "Only 1.3 billion worldwide employed full time for employer." August 12. Available at www.gallup.com/poll/174791/billion-worldwide-employed-full-time-employer.aspx

Gallup (2014b) "Worldwide, more men than women have full-time work." October 17. Available at www.gallup.com/poll/178637/worldwide-men-women-full-time-work.aspx.

Gautreaux, S.B. (2013) "Understanding China's internal migration." *International Policy Digest*, March 3. Available at http://intpolicy digest.org/2013/03/03/understanding-china-s-internal-migration.

Gialis, S., and Herod, A. (2014) "Of steel and strawberries: Greek workers struggle against informal and flexible working arrangements during the crisis." *Geoforum* 57: 138–149.

Gidda, M. (2016) "Jobs on the line: New technology could replace millions of call center workers in the Philippines." *Newsweek*, September 29. Available at www.newsweek.com/how-new-tech nology-automation-ai-take-away-millions-jobs-call-centers-503726.

Gilda Lehrman Institute of American History (n.d.) "Facts about the slave trade and slavery." Available at www.gilderlehrman.org/history-by-era/slavery-and-anti-slavery/resources/facts-about-slave-trade-and-slavery.

Gittins, J. (1981) *The Diggers from China: The Story of Chinese on the Goldfields.* Quartet Books: Melbourne.

Government of New South Wales (2010) "Australia's migration history." Available at www.migrationheritage.nsw.gov.au/belongings-home/about-belongings/australias-migration-history.

Guardian (2006) "Poles are bringing solidarity back into fashion in Britain." December 6.

Guardian (2012) "Nick Clegg pushes 'John Lewis'-style economy." January 15.

Guardian (2013) "Fashion chains sign accord to help finance safety in Bangladesh factories." May 13.

Guardian (2014) "Trafficked into slavery on Thai trawlers to catch food for prawns." June 10.

Guardian (2016) "Cambodians sue US and Thai firms over trafficking and forced labour claims." August 18.

Haines, M.R. (1994) "The population of the United States, 1790–1920." Historical Paper 56, National Bureau of Economic Research, Cambridge, MA.

Hannah, L. (1996) "Multinationality: Size of firm, size of country, and history dependence." *Business and Economic History* 25.2: 144–154.

Harington, J.S., McGlashan, N.D., and Chelkowska, E.Z. (2001) "A century of migrant labour in the gold mines of South Africa." *The Journal of the South African Institute of Mining and Metallurgy* 104.2: 65–71.

Harris, H.J. (1982) *The Right to Manage: Industrial Relations Policies of American Business in the 1940s.* University of Wisconsin Press: Madison.

Harris, J.E. (1971) *The African Presence in Asia: Consequences of the East African Slave Trade.* Northwestern University Press: Chicago, IL.

Harvey, D. (1982) *The Limits to Capital.* Blackwell: Oxford.

Harvey, D. (1989) *The Condition of Postmodernity: An Enquiry into the Origins of Cultural Change.* Blackwell: Oxford.

Heller, R. (2004) "Management styles in the new economy." *Management Issues,* March 4. Available at www.management-issues.com/opinion/6156/management-styles-in-the-new-economy.

Herod, A. (1998) "The geostrategics of labor in post-Cold War Eastern Europe: An examination of the activities of the International Metalworkers' Federation." In Herod, A. (ed.), *Organizing the Landscape: Geographical Perspectives on Labor Unionism.* University of Minnesota Press: Minneapolis, pp. 45–74.

Herod, A. (2000) "Implications of just-in-time production for union strategy: Lessons from the 1998 General Motors–United Auto Workers dispute." *Annals of the Association of American Geographers* 90.3: 521–547.

Herod, A. (2001) *Labor Geographies: Workers and the Landscapes of Capitalism.* Guilford Press: New York.

Herod, A. (2002) "Global change in the world of organized labor." In Johnston, R.J., Taylor, P.J., and Watts, M.J. (eds.), *Geographies of Global Change: Remapping the World,* 2nd edn. Blackwell: Oxford, pp. 78–87.

Herod, A. (2009) *Geographies of Globalization: A Critical Introduction.* Wiley-Blackwell: Chichester, UK.

Herod, A. (2011) "What does the 2011 Japanese tsunami tell us about the nature of the global economy?" *Social & Cultural Geography* 12.8: 829–837.

Herod, A., Pickren, G., Rainnie, A., and McGrath-Champ, S. (2014) "Global destruction networks, labour, and waste." *Journal of Economic Geography* 14.2: 421–441.

Hirst, P., and Thompson, G. (2001) *Globalization in Question.* Polity: Cambridge.

Hobsbawm, E.J. (1975) *The Age of Capital, 1848–1875.* Scribner: New York.

Hochschild, A.R. (2000) "Global care chains and emotional surplus value." In Giddens, A., and Hutton, W. (eds.), *On the Edge: Living with Global Capitalism.* Jonathan Cape: London, pp. 130–146.

Hodges, J., and Kahn, J. (2016) "UK Uber drivers' court victory over pay could rejigger gig economy." Available at www.insurancejournal.com/news/international/2016/10/31/430876.htm.

Holman, D., Batt, R., and Holtgrewe, U. (2007) *The Global Call Center Report: International Perspectives on Management and Employment (Executive Summary).* Cornell University School of Industrial and Labor Relations: Ithaca, NY.

Hossain, M., and Islam, M. (2006) *Ship Breaking Activities and its Impact on the Coastal Zone of Chittagong, Bangladesh: Towards Sustainable Management.* Young Power in Social Action (YPSA):

Chittagong, Bangladesh. Available at www.shipbreakingbd.info/report/Ship%20Breaking%20Activities%20and%20its%20Impact%20on%20the.pdf.

Howard, A. (1995) "Global capital and labor internationalism in comparative historical perspective: A Marxist analysis." *Sociological Inquiry* 65.3–4: 365–394.

Hoxhaj, R., Marchal, L., and Seric, A. (2015) "FDI and migration of skilled workers towards developing countries: Firm-level evidence from Sub-Saharan Africa." *Journal of African Economies* 25.2: 201–232.

ILO (2010) *Decent Work for Domestic Workers: Report IV(1)*. International Labour Organization: Geneva.

ILO (2011) *Global and Regional Estimates on Domestic Workers*. International Labour Organization: Geneva.

ILO (2012) *From Precarious Work to Decent Work: Outcome Document to the Workers' Symposium on Policies and Regulations to Combat Precarious Employment*. International Labour Organization, Bureau for Workers' Activities: Geneva.

ILO (2013a) *Marking Progress Against Child Labour: Global Estimates and Trends 2000–2012*. International Labour Organization, Governance and Tripartism Department: Geneva.

ILO (2013b) *Domestic Workers Across the World: Global and Regional Statistics and the Extent of Legal Protection*. International Labour Organization: Geneva.

ILO (2015) *World Employment and Social Outlook: The Changing Nature of Jobs*. International Labour Organization: Geneva.

ILO (2016) *ASEAN in Transformation: How Technology is Changing Jobs and Enterprises*. International Labour Organization, Bureau for Employers' Activities: Geneva.

Infinit Contact (2014) "Why is India losing 70% of call center business to the Philippines?" May 7. Available at www.infinitcontact.com/blog/india-losing-70-call-center-business-philippines.

International Business Times (2014) "Samsung apology rekindles questions over safety of computer chip manufacturing." May 14.

International Cocoa Organization (2014) *The Cocoa Market Situation*. Report by the Economics Committee, dated September 16–18. Available at www.icco.org/about-us/international-cocoa-agreements/doc_download/1383-cocoa-market-situation-24-july-2014.html.

Intuit (2015) "7.6 million people in on-demand economy by 2020." Available at www.businesswire.com/news/home/20150813005317/

en/Intuit-Forecast-7.6-Million-People-On-Demand-Economy#.Ve
dGWE3bJon.

IOM (2014a) *Global Migration Trends: An Overview.* International
Organization for Migration: Geneva.

IOM (2014b) *Notes on Migration and Development in the Global South:
Emerging Issues and Responses.* International Organization for
Migration: Geneva.

IOM (2015) *Dinámicas migratorias en América Latina y el Caribe
(ALC), y entre ALC y la Unión Europea.* International Organization
for Migration: Brussels.

Irish Times (2009) "80 new jobs for Belfast call centre." August 7.

Ishemo, S.-L. (1995) "Forced labour and migration in Portugal's
African colonies." In Cohen, R. (ed.), *The Cambridge Survey
of World Migration.* Cambridge University Press: Cambridge,
pp. 162–165.

ITUC (2011) *Labour Standards in World Bank Group Lending: Lessons
Learned and Next Steps.* Report dated November. International
Trade Union Confederation: Brussels.

ITUC (2014) *The Case Against Qatar, Host of the FIFA 2022 World Cup.*
Report dated March. International Trade Union Confederation:
Brussels.

ITUC (2016) "About us." Available at www.ituc-csi.org/about-us.

Jeon, J. (2014a) "More than 2 million in S. Korea living as subcontrac-
tor workers." *The Hankyoreh* (South Korea). September 29.

Jeon, J. (2014b) "With irregular jobs, S. Korea risks another Sewol
every day." *The Hankyoreh* (South Korea). May 26.

Johns, R. (1998) "Bridging the gap between class and space: US
worker solidarity with Guatemala." *Economic Geography* 74.3:
252–271.

Johns, R., and Vural, L. (2000) "Class, geography, and the consumer-
ist turn: UNITE and the Stop Sweatshops Campaign." *Environment
and Planning A* 32.7: 1193–1213.

Jones, G. (1996) *The Evolution of International Business: An Introduction.*
Routledge: London.

Jones, G. (2000) *Merchants to Multinationals: British Trading Companies
in the Nineteenth and Twentieth Centuries.* Oxford University Press:
Oxford.

Jones, G. (2005) *Multinationals and Global Capitalism: From the
Nineteenth to the Twenty-First Century.* Oxford University Press:
Oxford.

Katz, C. (2004) *Growing Up Global: Economic Restructuring and Children's Everyday Lives*. University of Minnesota Press: Minneapolis.

Kessler, S. (2015) "The gig economy won't last because it's being sued to death." *Fast Company*, February 17.

Kumar, D. (1983) *The Cambridge Economic History of India, Volume 2: c.1757–c.1970*. Cambridge University Press: Cambridge.

Lavaca Collective (2007) *Sin Patrón: Stories from Argentina's Worker-Run Factories*. Haymarket Books: Chicago, IL.

Lazonick, W. (2009) *Sustainable Prosperity in the New Economy? Business Organization and High-Tech Employment in the United States*. W.E. Upjohn Institute for Employment Research: Kalamazoo, MI.

Lewis, J.S. (1948) *Partnership for All: A Thirty-Four-Year-Old Experiment in Industrial Democracy*. Kerr-Cros Publishing: London.

Library of Congress (n.d.) "Hawaii: Life in a plantation society." Available at www.loc.gov/teachers/classroommaterials/presentations andactivities/presentations/immigration/japanese2.html.

Lier, D. (2007) "Places of work, scales of organising: A review of labour geography." *Geography Compass* 1.4: 814–833.

Looney, R. (2003) "The neoliberal model's planned role in Iraq's economic transition." *Middle East Journal* 57.4: 568–586.

Los Angeles Times (1992) "Computer chip workers risk miscarriages." December 4.

Los Angeles Times (2016) "Judge rejects Uber's $100-million class-action settlement, calling it unfair to drivers." August 18.

Luce, S. (2004) *Fighting for a Living Wage*. Cornell University Press: Ithaca, NY.

Lynch, K.A. (2003) *Individuals, Families, and Communities in Europe, 1200–1800*. Cambridge University Press: Cambridge.

Magtulis, P. (2012) "Remittance growth poised to meet full-year forecast – BSP." *The Philippine Star*, November 15.

Manchester, A.K. (1964) *British Preëminence in Brazil: Its Rise and Decline: A Study in European Expansion*. Octagon Books: New York.

Mantouvalou, V. (2012) "Human rights for precarious workers: The legislative precariousness of domestic labour." *Comparative Labor Law and Policy Journal* 34.1: 133–165.

Maritime Union of New Zealand (2014) "Maritime workers recover crews' wages in New Zealand ITF week of action." Media release, July 7. Available at www.munz.org.nz/2014/07/07/maritime-workers-recover-crews-wages-in-new-zealand-itf-week-of-action/.

Marks, B. (2012) "The political economy of household commodity production in the Louisiana shrimp fishery." *Journal of Agrarian Change* 12.2 and 3: 227–251.

Marx, K. (1852 [1963]) *The Eighteenth Brumaire of Louis Bonaparte*. International Publishers: New York (2004 printing).

Marx, K. (1867 [1967]) *Capital: A Critique of Political Economy, Volume One*. International Publishers: New York (1987 printing).

Mattingly, D.J. (1999) "Job search, social networks, and local labor-market dynamics: The case of paid household work in San Diego, California." *Urban Geography* 20.1: 46–74.

Mavroudeas, S.D. (2013) "Development and crisis: The turbulent course of Greek capitalism." *International Critical Thought* 3.3: 297–314.

M'bokolo, E. (1998) "The impact of the slave trade on Africa." *Le Monde diplomatique*. April.

McCarthy, N. (2015) "The world's biggest chocolate consumers." *Forbes* on-line. Available at www.forbes.com/sites/niallmccarthy/2015/07/22/the-worlds-biggest-chocolate-consumers-infographic/#1c841f5312b8.

McKinsey and Company (2015) *A Labor Market That Works: Connecting Talent with Opportunity in the Digital Age*. McKinsey Global Institute, June.

Merk, J. (2009) *Stitching a Decent Wage Across Borders: The Asia Floor Wage Proposal 2009*. Report from the Clean Clothes Campaign on behalf of the Asia Floor Wage Campaign, New Delhi.

Min, P.-G. (1992) "A comparison of the Korean minorities in China and Japan." *International Migration Review* 26.1: 4–21.

Mitchell, D. (2012) *They Saved the Crops: Labor, Landscape, and the Struggle over Industrial Farming in Bracero-Era California*. University of Georgia Press: Athens.

Morris, R. (2012) *Scoping Study: Impact of Fly-in Fly-out/Drive-in Drive-out Work Practices on Local Government*. Australian Centre of Excellence for Local Government, University of Technology, Sydney.

NAPS (2016) "Why Mexico's electronics manufacturing is growing." NAPS (North American Production Sharing, Inc.), Solana Beach, CA, February 23.

NASSCOM-McKinsey (2005) Nasscom-McKinsey 2005 Report: Extending India's Leadership of the Global IT and BPO Industries. NASSCOM-McKinsey: New Delhi.

Nest, M. (2011) *Coltan*. Polity: Cambridge.

New York Times (1996) "Ailing computer-chip workers blame chemicals, not chance." March 28.

New York Times (2000) "TECHNOLOGY; Temp workers at Microsoft win lawsuit." December 13.

New York Times (2002) "In New York tickets, Ghana sees orderly city." July 22.

New York Times (2004) "Canada, the closer country for outsourcing work." November 30.

New York Times (2006) "The long-distance journey of a fast-food order." April 11.

New York Times (2008) "The food chain: Environmental cost of shipping groceries around the world." April 26.

New York Times (2012) "Trade union group, ANC ally, holds strikes in South Africa." March 7.

NGA (2016) *Raising the Floor for Supply Chain Workers: Perspective from U.S. Seafood Supply Chains*. National Guestworker Alliance: New Orleans, LA.

Northrup, D. (1999) "Migration from Africa, Asia, and the South Pacific." In Porter, A. (ed.), *The Oxford History of the British Empire*, vol. 3. Oxford University Press: Oxford, pp. 88–100.

Northrup, D. (2000) "Indentured Indians in the French Antilles. Les immigrants indiens engagés aux Antilles françaises." *Revue française d'histoire d'outre-mer* 87.326: 245–271.

OECD (2012) *OECD Employment Outlook 2012*. Organisation for Economic Cooperation and Development: Paris.

OECD (2016a) *Labour Force Statistics 2005–2014*. Organisation for Economic Cooperation and Development: Paris.

OECD (2016b) Employment Database – Employment Indicators. Available at www.oecd.org/employment/emp/employmentdatabase-employment.htm.

Officer, L.H., and Williamson, S.H. (2016) "Five ways to compute the relative value of a UK pound amount, 1270 to present." Available at www.measuringworth.com/ukcompare/.

Ohmae, K. (2005) *The Next Global Stage: Challenges and Opportunities in Our Borderless World*. Wharton School Publishing: Upper Saddle River, NJ.

Ormaechea, J.M. (2003) *Medio siglo de la experiencia cooperativa de Mondragón*. Otalora: Mondragón.

Parliament of Western Australia (2015) *The Impact of FIFO Work Practices on Mental Health: Final Report*. Report by the Education and Health Standing Committee, Legislative Assembly, Parliament of Western Australia, Perth.

Parreñas, R.S. (2001) *Servants of Globalization: Women, Migration and Domestic Work*. Stanford University Press: Stanford, CA.

Peet, R. (1983) "Relations of production and the relocation of United States manufacturing industry since 1960." *Economic Geography* 59.2: 112–143.

Pérotin, V. (2016) *What Do We Really Know About Worker Co-operatives?* Report for Co-operatives UK Limited, Manchester.

Piore, M.J., and Sabel, C.F. (1984) *The Second Industrial Divide: Possibilities for Prosperity*. Basic Books: New York.

Plowman, D.H., Deery, S., and Fisher, C. (1980) *Australian Industrial Relations*, rev. edn. McGraw-Hill: Sydney.

Poster, W.R. (2007) "Who's on the line? Indian call center agents pose as Americans for US-outsourced firms." *Industrial Relations* 46.2: 271–304.

Pradhan, J.K., and Abraham, V. (2005) "Social and cultural impact of outsourcing: Emerging issues from Indian call centers." *Harvard Asia Quarterly* 9.3: 22–30.

Pratt, G. (1999) "From registered nurse to registered nanny: Discursive geographics of Filipina domestic workers in Vancouver, BC." *Economic Geography* 75.3: 215–236.

Rainnie, A., Herod, A., and McGrath-Champ, S. (2013) "Global production networks, labour and small firms." *Capital and Class* 37.2: 177–195.

Ramstad, E. (1994) "Intel to demonstrate next generation chip within a year." *AP News Archive*, January 27. Available at www.apnews archive.com/1994/Intel-to-Demonstrate-Next-Generation-Chip-Within-a-Year/id-5c180eea20a7dde3aab1e9322e20800b.

Ranis, P. (2005) "Argentina's worker-occupied factories and enterprises." *Socialism and Democracy* 19.3: 93–115.

Rapelli, S. (2012) *European I-Pros: A Study*. Report for the European Forum of Independent Professionals, Brussels. Available at www.efip.org/the-growth-of-european-independent-professionals.

Rio Tinto (2014) *Rio Tinto: Mine of the Future™*. Rio Tinto plc: London.

Rodis, R. (2013) "Why are there so many Filipino nurses in the US?" May 12. *Inquirer.net* (Philippines). Available at http://globalnation.inquirer.net/74321/why-are-there-so-many-filipino-nurses-in-the-us.

Rutherford, T. (2010) "De/re-centering work and class? A review and critique of labour geography." *Geography Compass* 4.7: 768–777.

San Francisco Chronicle (2012) "Samsung-Apple trial hears of sacrifice." August 15.

SCA (2012) "It's tough work for our cleaners." *European Cleaning Journal*, November 2. Available at www.europeancleaningjourn al.com/magazine/articles/case-studies/its-tough-work-for-our-cleaners.

Schooling, J.H. (1911) *The British Trade Book*. J. Murray: London.

Scott, J.C. (1985) *Weapons of the Weak: Everyday Forms of Peasant Resistance*. Yale University Press: New Haven, CT.

Scott, R.E. (2015) "The manufacturing footprint and the importance of U.S. manufacturing jobs." *Economic Policy Institute Briefing Paper #388*. Available at www.epi.org/publication/the-manufacturing-footprint-and-the-importance-of-u-s-manufacturing-jobs.

Seafarers' International Union (2015) "ITF recovers $59.5 million for mariners in Europe, Asia." Seafarers' Log Media release, September. Available at www.seafarers.org/seafarerslog/2015/September2015/ITFRecoversWages.htm.

Segal, A. (1993) *An Atlas of International Migration*. Hans Zell Publishers: London.

Shanghai Daily (2013) "Intel to make Chengdu a global supply center." June 6.

Shaw, C. (2005) "Rothschilds and Brazil: An introduction to sources in the Rothschild archive." *Latin American Research Review* 40.1: 165–185.

Sherry, D. (2010) *Occupy! A Short History of Workers' Occupations*. Bookmarks: London.

Silver, B.J. (2003) *Forces of Labor: Workers' Movements and Globalization Since 1870*. Cambridge University Press: Cambridge.

Skeldon, R. (2011) "China: An emerging destination for economic migration." Migration Policy Institute, May 31. Available at www.migrationpolicy.org/article/china-emerging-destination-economic-migration.

Solis, H. (2012) "Providing protections for in-home care workers." Posting on the Official Blog of the US Department of Labor. Available at https://obamawhitehouse.archives.gov/blog/2011/12/15/provid ing-protections-home-care-workers.

South China Morning Post (2010) "Foxconn factories are labour camps: Report." October 11.

Southall, H. (1988) "Towards a geography of unionization: The spatial organization and distribution of early British trade unions."

Transactions of the Institute of British Geographers, New Series 13.4: 466–483.

Southall, H. (1989) "British artisan unions in the New World." *Journal of Historical Geography* 15.2: 163–182.

Standing, G. (2011) *The Precariat: The New Dangerous Class.* Bloomsbury Academic: London.

Stern, A. (2006) *A Country That Works: Getting America Back on Track.* Free Press: New York.

Stigler, G.J. (1946) "The number of servants, 1900–1940." In Stigler, G.J. (ed.), *Domestic Servants in the United States, 1900–1940.* National Bureau of Economic Research: New York, pp. 1–6.

Storey, K. (2001) "Fly-in/Fly-out and Fly-over: Mining and regional development in Western Australia." *Australian Geographer* 32.2: 133–148.

Storper, M., and Walker, R. (1983) "The theory of labour and the theory of location." *International Journal of Urban and Regional Research* 7.1: 1–42.

Storper, M., and Walker, R. (1989) *The Capitalist Imperative: Territory, Technology, and Industrial Growth.* Blackwell: Malden, MA.

Stover, J., and Bollinger, L., (1999) "The economic impact of AIDS." Research Paper for The Futures Group International, in Collaboration with the Research Triangle Institute (RTI) and The Centre for Development and Population Activities (CEDPA).

Stringer, C., Whittaker, D.H., and Simmons, G. (2016) "New Zealand's turbulent waters: The use of forced labour in the fishing industry." *Global Networks: A Journal of Transnational Affairs* 16.1: 3–24.

Sydney Morning Herald (2012) "Fly-in fly-out saves millions, Fortescue tells inquiry." April 18.

Sydney Morning Herald (2015) "Some FIFO workers pay high price for riches." January 11.

Taylor, P. and Bain, P. (2006) *An Investigation into the Offshoring of Financial Services Business Processes.* A Report for Scottish Enterprise – Scottish Development International: Department of Human Resource Management, University of Strathclyde, Glasgow.

Taylor, P. and Bain, P. (2010) "'Across the great divide': Local and global trade union responses to call centre offshoring to India." In McGrath-Champ, S., Herod, A., and Rainnie, A. (eds.), *Handbook of Employment and Society: Working Space.* Edward Elgar: Cheltenham, pp. 436–456.

Telegraph (2001) "The child slaves of the Ivory Coast – bought and sold for as little as £40." April 22.

The Age (2014) "Student by day, office cleaner by night: International students lead fight for better pay." July 28. Available at www.theage. com.au/national/education/student-by-day-office-cleaner-by-night-international-students-lead-fight-for-better-pay-20140718-zujpc.html.

The Economist (2016) "Call centres: The end of the line." February 6. Available at www.economist.com/news/international/21690041-call-centres-have-created-millions-good-jobs-emerging-world-technology-threatens.

Thomas, M. (2016) "Global unions, global framework agreements and the transnational regulation of labour standards." In Lambert, R., and Herod, A. (eds.), *Neoliberal Capitalism and Precarious Work: Ethnographies of Accommodation and Resistance*. Edward Elgar: Cheltenham, pp. 277–302.

Thompson, E.P. (1963) *The Making of the English Working Class*. Vintage Books: New York.

Tijuana EDC (n.d.) *Tijuana: The Electronics Sector Nearshore Manufacturing Capital*. Tijuana Economic Development Corporation: Tijuana, Mexico.

Tulane University (2015) *Final Report 2013/14: Survey Research on Child Labor in West African Cocoa Growing Areas*. School of Public Health and Tropical Medicine, Tulane University, New Orleans, LA.

UK Office for National Statistics (2014) *Statistical Bulletin – Annual Survey of Hours and Earnings: 2014 Provisional Results*. Office for National Statistics: Newport, UK.

UNCTAD (2006) *World Investment Report 2006*. United Nations Conference on Trade and Development: Geneva.

UNCTAD (2015a) *World Investment Report 2015*. United Nations Conference on Trade and Development: Geneva.

UNCTAD (2015b) *Review of Maritime Transport 2015*. United Nations Conference on Trade and Development: Geneva.

UNCTAD (2016a) *World Investment Report 2016*. United Nations Conference on Trade and Development: Geneva.

UNCTAD (2016b) *World Investment Report 2016: Annex Tables*. United Nations Conference on Trade and Development: Geneva. Available at http://unctad.org/en/Pages/DIAE/World%20Investment%20 Report/Annex-Tables.aspx.

UNFPA (2007) *State of World Population 2007: Unleashing the Potential of Urban Growth*. United Nations Population Fund: New York.

United Nations (1998) *Briefing Packet, 1998 Revision of World Population Prospects*. United Nations Department of Economic and Social Affairs, Population Division: New York.

United Nations (2004) *World Population to 2300*. United Nations Department of Economic and Social Affairs, Population Division: New York.

United Nations (2012) *Migrants by Origin and Destination: The Role of South-South Migration*. United Nations Department of Economic and Social Affairs, Population Division: New York.

United Nations (2013) *International Migration Report 2013*. United Nations Department of Economic and Social Affairs, Population Division: New York.

United Nations (2014) *World Urbanization Prospects*. United Nations: New York.

United Nations (2015a) *World Population Prospects: The 2015 Revision*. United Nations Department of Economic and Social Affairs, Population Division: New York. Online version.

United Nations (2015b) *World Population Prospects: The 2015 Revision – Key Findings and Advance Tables*. United Nations Department of Economic and Social Affairs, Population Division: New York.

United Nations (2016) *International Migration Report 2015: Highlights*. United Nations Department of Economic and Social Affairs, Population Division: New York.

Upadhyay, P., Khokhar, A., and Talwar, R. (2012) "A cross-sectional study on the level of perceived stress and self-reported morbidity among call handlers working in call centres in Delhi NCR." Paper presented at the 9th Joint Annual Conference of the Indian Society for Malaria & Other Communicable Diseases, November 2–4, New Delhi.

US Bureau of Labor Statistics (2012) "May 2012 national industry-specific occupational employment and wage estimates: NAICS 561420 – Telephone call centers." Available at www.bls.gov/oes/2012/may/naics5_561420.htm.

US Bureau of Labor Statistics (2015a) "Labor force statistics from the current population survey: (Seas) employed, usually work part time." Available at http://data.bls.gov/timeseries/LNS12600000?series_id=LNS12000000.

US Bureau of Labor Statistics (2015b) *Occupational Outlook Handbook, 2016–17 Edition. Janitors and Building Cleaners.* Available at www.bls.gov/ooh/building-and-grounds-cleaning/janitors-and-building-cleaners.htm.

US Bureau of Labor Statistics (2016) "Nonfatal occupational injuries and illnesses requiring days away from work, 2015." Press release dated November 10.

US Government Accountability Office (2006) *Employment Arrangements: Improved Outreach Could Help Ensure Proper Worker Classification.* Report GAO-06-656 to the Ranking Minority Member, Committee on Health, Education, Labor, and Pensions, US Senate, Washington, DC.

US Government Accountability Office (2011) *Seafood Safety: FDA Needs to Improve Oversight of Imported Seafood and Better Leverage Limited Resources.* GAO: Washington, DC.

US Government Accountability Office (2015) *Contingent Workforce: Size, Characteristics, Earnings, and Benefits.* Report GAO-15-168R to US Senators Murray and Gillibrand, US Senate, Washington, DC.

US State Department (2016) *Trafficking in Persons Report.* June. Office of the Under Secretary for Civilian Security, Democracy, and Human Rights, US State Department: Washington, DC.

Vaid, M. (2009) *Exploring the Lives of Youth in the BPO Sector: Findings from a Study in Gurgaon.* Health and Population Innovation Programme Working Paper, No. 10, Population Council, New Delhi, India.

van der Linden, M. (1988) "The rise and fall of the First International: An interpretation." In van Holthoon, F. and van der Linden, M. (eds.), *Internationalism in the Labour Movement 1830–1940, Volume 1.* E.J. Brill: London, pp. 323–335.

van Holthoon, F., and van der Linden, M. (eds.) (1988) *Internationalism in the Labour Movement 1830–1940, Volumes 1 and 2.* E.J. Brill: London.

Vogel, S. K. (1996) *Freer Markets, More Rules: Regulatory Reform in Advanced Industrial Countries.* Cornell University Press: Ithaca, NY.

Wainwright, H., and Elliott, D. (1982) *The Lucas Plan: A New Trade Unionism in the Making?* Allison & Busby: London.

Wall Street Journal (2016) "Robots on track to bump humans from call-center jobs." June 21.

Weber, A. (1909 [1929]) *Über den Standort der Industrie.* J.C.B. Mohr (Paul Siebeck): Tübingen (published in English in 1929 as *Theory of the Location of Industries* by the University of Chicago Press, Chicago).

Western Australian Local Government Association (2010) *Research into the Impacts of Fly-in/Fly-out on Western Australian Communities: Literature Review.* Submitted to the Western Australia Parliamentary inquiry into FIFO/DIDO work practices.

Whyte, W.F., and Whyte, K.K. (1991) *Making Mondragón: The Growth and Dynamics of the Worker Cooperative Complex,* 2nd edn. Cornell University Press: Ithaca, NY.

Wilkins, M. (1994) "Comparative hosts." *Business History* 36.1: 18–50.

Wills, J. (2002) "Bargaining for the space to organize in the global economy: A review of the Accor–IUF trade union rights agreement." *Review of International Political Economy* 9.4: 675–700.

Wills, J., and Linneker, B. (2012) *The Costs and Benefits of the London Living Wage.* Report for the Trust for London, October.

Winn, P. (2014) "How the Philippines is crushing the Indian call center business." Public Radio International, May 4. Available at www. pri.org/stories/2014-05-04/how-philippines-crushing-indian-call-center-business.

Wittman, J.S., Jr. (1973) *Selected Articles in Social Ecology.* MSS Information Corporation: New York.

Wolff, R.D., and Resnick, S.A. (1987) *Economics: Marxian Versus Neoclassical.* Johns Hopkins Press: Baltimore, MD.

Woodruff, W. (1966) *Impact of Western Man: A Study of Europe's Role in the World Economy, 1750–1960.* St. Martin's Press: New York.

World Cocoa Foundation (2014) "Cocoa market update." April 1. Available at www.worldcocoafoundation.org/wp-content/uploads/Cocoa-Market-Update-as-of-4-1-2014.pdf.

Wright, E.O. (2000) "Working-class power, capitalist-class interests, and class compromise." *American Journal of Sociology* 105.4: 957–1002.

Xiang, B. (2016) *Emigration Trends and Policies in China: Movement of the Wealthy and Highly Skilled.* Migration Policy Institute: Washington, DC.

Yannopoulos, G.N. (1988) "The economics of 'flagging out'." *Journal of Transport Economics and Policy* 22.2: 197–207.

Yeung, H. (2009) "Transnational corporations, global production networks, and urban and regional development: A geographer's perspective on 'Multinational Enterprises and the Global Economy'." *Growth and Change* 40.2: 197–226.

Yun, A. (2016) "Organizing across a fragmented labour force: Trade union responses to precarious work in Korean auto companies." In Lambert, R., and Herod, A. (eds.), *Neoliberal Capitalism and Precarious Work: Ethnographies of Accommodation and Resistance.* Edward Elgar: Cheltenham, UK, pp. 201–225.

Index

Grove, Andrew (Intel
Corporation), 141
Gulf Cooperation Council
(GCC), 34, 203 n.2:2

Harvey, David, 20, 91
HIV/AIDS, impact of on labor
force development, 52–3
Howard, John, 116
Hyundai, labor relations at,
188

IG Metall (Germany), 181, 182
illegal drugs, use of by workers,
133
India (*passim*), 36, 60, 61, 67,
77, 82, 92, 114, 120, 145,
150, 175, 178, 188
commercial cleaning sector,
162, 182
impact of British imperialism,
27, 49, 61–2
migrant workers from, 3, 30,
38, 39, 40, 43, 54, 55
population dynamics, 49–52
shipbreaking industry, 74–5,
78
see also call center industry:
India, development of in
industrialization
global food production,
impact of on, 27–8, 59
Global North population
dynamics, impact of on,
25–8, 48–50
in the Global South, 59, 61,
102, 107
import-substitution
industrialization, 64
IndustriALL, see global union
federations
Institute for Global Labour and
Human Rights (formerly

National Labor Committee)
(US), 187
International Association, 174
International Confederation
of Free Trade Unions
(ICFTU), 176, 181
International Federation of
Christian Trade Unions,
176–8
International Federation of
Trade Unions, 176
International Framework
Agreements, 180
Accor hotel chain and, 180
International Ladies' Garment
Workers' Union, 181
International Metalworkers'
Federation, 179
see also global union
federations
International Research Network
on Autowork in the
Americas, 194
International Secretariat of
National Trade Union
Centres, 176
International Trade Union
Confederation (ITUC), 144,
178, 179, 185
World Day for Decent Work,
185–6
International Transport
Workers' Federation (ITF),
130, 178
see also global union
federations
International Union of Food
and Allied Workers'
Associations, 195
see also global union
federations
International Workingmen's
Association, 174